Reality Reloaded

The Scientific Case for a Simulated Universe

by

Melvin M. Vopson

Associate Professor of Physics,
University of Portsmouth

Research funded by

Information Physics Institute

Published by

IPI Publishing
Hampshire, UK
2023

Information Physics Institute
IPI Publishing
Gosport, Hampshire
United Kingdom
https://www.informationphysicsinstitute.org/

M.M. Vopson, Reality Reloaded: The Scientific Case for a Simulated Universe, IPI Publishing, ISBN **978-1-80517-057-0**, (2023) DOI: *https://doi.org/10.59973/rrtscfasu*

ISBN **978-1-80517-057-0** (print)

ISBN **978-1-80517-058-7** (ebook)

Table of Contents

Preface   **v**

1   **Digital Big Data:**
      A Hint Why the Universe Might Be a Simulation   **1**

     1.1   The Information Catastrophe   3
     1.2   Eddington number   7

2   **Physical Facts**
      Supporting the Simulated Universe Hypothesis   **11**

     2.1   Pixalation of the universe   14
     2.2   The universe is mathematical   15
     2.3   Speed of light   16
     2.4   Quantum entanglement   17
     2.5   Wave-particle duality   19
     2.6   The Big Bang and the inflation theory   20
     2.7   Pauli's exclusion principle   22
     2.8   Cambrian explosion. Where did all the information
         come from?   24

3   **Brief Introduction to Shannon's Information Entropy**   **27**

     3.1   Generating a set of m-blocks   32
     3.2   Calculating the IE of a digital file   33
     3.3   Calculating the IE of a genomic sequence   34

4   **Information is the Fifth State of Matter**   **39**

     4.1   Entropy of the information bearing states   40
     4.2   Landauer's principle   44
     4.3   Global information power estimation   47
     4.4   Mass – Energy – Information equivalence principle   49
     4.5   The mass of the world's data   54
     4.6   Is dark matter information?   56
     4.7   The DNA of matter   59
     4.8   The information content per elementary particle   62

5   Proposed Experiments for Testing the Information Conjectures   **67**

  5.1 Mass of data   67
  5.2 Mass of hot objects   68
  5.3 Detection of information at erasure   70
     5.3.1 Theoretical predictions   74
     5.3.2 Proposed experimental design   75

6   Second Law of Information Dynamics   **81**

  6.1 The necessity of the second law of infodynamics   82
  6.2 Second law of infodynamics and digital information   87
  6.3 Second law of infodynamics and genetic information   90
  6.4 Second law of infodynamics and Hund's Rule   95
     6.4.1 Numerical calculations   98
     6.4.2 s – orbital   100
     6.4.3 p – orbital   100
     6.4.4 d – orbital   102
     6.4.5 f – orbital   103
  6.5 Second law of infodynamics and symmetries   104

7   A Religious Perspective to the Simulated Universe Theory   **113**

8   Concluding remarks   **117**

Appendix A The abundance of elements in the universe   **123**

References   **137**

Acknowledgements   **145**

Preface

The book "*Reality Reloaded: The Scientific Case for a Simulated Universe*" is an expanded collection of the author's own scientific articles published in the past five years on the topic of information physics. The book caters to a diverse audience, including scientists, academics, students, and the general public. The book is written in a scientific style, yet it is accessible by making complex concepts easy to understand and by ensuring engagement for both experts and those new to the subject.

At its core, the simulated universe hypothesis suggests that the universe we inhabit, with all its galaxies, stars, planets, and life forms, is a meticulously programmed simulation. In this scenario, the physical laws governing our reality are algorithms, and the tangible experiences we have are simply generated by the computational processes of an immensely advanced system. While the book does not question who might be responsible for running the simulation, it could be run by a post-human civilization, an extraterrestrial intelligence, or even a future iteration of ourselves.

While inherently speculative, the simulated universe theory has garnered attention from scientists, philosophers, and even popular culture due to its intriguing implications. The idea has made its mark in popular culture, influencing movies, TV shows, and books. The 1999 film "The Matrix," directed by the Wachowskis, famously depicted a world where humans exist within a simulated reality. This portrayal popularised the idea and introduced it to mainstream audiences.

Proponents of the simulated universe hypothesis offer several arguments in support of its plausibility, although it's important to note that these arguments remain speculative and lack empirical evidence. One key argument is the exponential growth of computing power. As our own technology progresses, we are approaching the point where we could simulate complex virtual worlds ourselves. If this trend continues, it suggests that a sufficiently advanced civilization could also achieve the capability to create simulations that are indistinguishable from reality.

Another argument draws from the principles of simulation theory in computer science and gaming. Our own simulations, such as video games and virtual reality environments, are becoming increasingly sophisticated and immersive. If our own creations are progressing in this direction, it follows that a more advanced civilization could create simulations that are vastly superior in fidelity and complexity.

Within the scientific community, the concept of a simulated universe has sparked both fascination and scepticism. Some scientists suggest that if our reality is a simulation, there may be glitches or patterns within the fabric of the universe that betray its simulated nature. However, the search for such anomalies remains a challenge, as our understanding of the laws of physics is still evolving, and we lack a definitive framework to distinguish between a simulated and non-simulated reality.

The key question remains: *Can we find scientific evidence to support or refute the simulated universe hypothesis?*

Answering this key question is the main focus of this book. The author delves into the connections between information science, technological advancements, cutting-edge concepts in physics, and the plausibility of the simulated universe hypothesis, offering unique perspectives and novel scientific arguments to the reader.

The author demonstrates scientifically that there is empirical evidence all around us showing that the information content associated with any event or process in the universe is always minimized. This behaviour is reminiscent of the rules deployed in programming languages and computer coding. The empirically observed effect is in fact a built-in universal data optimisation and compression mechanism, which supports the idea of a simulated universe because it requires a reduction in computational power and data storage to run the simulation.

While the facts presented here cannot be regarded as ultimate proof that the simulated universe hypothesis is true, the book is very stimulating and invites further research in the fascinating field of information physics.

1

Digital Big Data: A Hint Why the Universe Might Be a Simulation

Information storage is essential for a society to advance and preserve knowledge, history, the arts and culture. Ancient humans stored information in cave paintings, the oldest we know of are over 40,000 years old. As humans evolved, the emergence of languages and the invention of writing led to detailed information being stored in various written forms, culminating with the invention of paper in China around the first century AD. The oldest printed books appeared in China between AD 600 and AD 900. For over a millennium, books remained the main source of information storage.

Fast forward to our modern times, and humans have achieved more technological development in the past 150 years than during the previous 5,000 years. Arguably some of the most important developments in human history were the discovery of electricity generation by Faraday in 1831 and the emergence of digital electronics. Since the first discovery of the transistor in 1947 and the integrated microchip in 1956, our society has undergone huge technological developments. In just over half a century since the beginning of the Silicon revolution, we have achieved unprecedented computing power, wireless technologies, the Internet, artificial intelligence, and multiple technological advances in display technologies, mobile communications,

1

transportation and medicine, to name a few. However, none of these could have been possible without mastering the ability to create and store large amounts of digital information.

IBM developed the first digital magnetic data storage hard disc drive in 1956. The memory device had a storage capacity of just under 5 Mb and it was first used in the IBM 305 RAMAC system.

Today's data storage devices are cost-effective and have Tb of storage capacity. In fact, digital information is a valuable commodity and the backbone of some of the largest hi-tech companies in the world today. Most importantly, the introduction of digital data storage also changed the way we produce, manipulate, and store information. The transition point took place in 1996, when digital storage became more cost-effective for storing information than paper[1].

A rich variety of digital data storage technologies are available today, and they could be classified into three main categories: a) magnetic storage technologies (HDD, tape); b) optical storage technologies (CD, DVD, Blu-Ray); c) semiconductor memories (SSD, flash drive).

Each type of memory is useful for specific applications. For example, semiconductor memories are the preferred choice for portable electronics; optical storage is mostly used for movies, software and gaming; and magnetic data storage remains the dominant technology for high-capacity information storage, including personal computers and data servers.

Regardless of the type of data storage technology, the operating principles are the same. Digital bits of information can be stored in any material containing two distinctive and switchable physical states. In binary code, the digital information is stored as ones and zeroes, also known as bits. Eight bits form a byte. A logical zero or one is allocated to each physical state. The smaller these physical states are, the more bits can be packed into the storage device. The size of a single digital bit today is on the order of tens of nanometres (billionths of a metre). These devices are very complex because developing devices capable of storing information at this scale requires controlling materials on the atomic level.

1.1 The Information Catastrophe

Future	The Information Catastrophe
. . .	. . .
2020	World creates 2.5 billion Gb of information every day
2019	5G wireless communications launched
2018	Virtual Cloud Network was launched
2013	Fast Data era- YouTube hits 1 billion users
2012	Facebook hits 1 billion users
2007	The first generation iPhone (smart phone from Apple) was released
2006	Twitter service was launched
2005	YouTube was launched
2004	Facebook was launched
1998	Google was founded
1996	Digital storage becomes cheaper for storing data than paper (IBM Systems Journal, 2003)
1983	TCP/IP protocol developed - birth of Internet
1970s	First PCs enter consumer markets
1969	ARPAnet delivers first communication from one computer to another
1958	First microchip developed at Texas Instruments
1956	IBM develops first HDD
1947	First transistor developed at Bell laboratories

Figure 1.1

3

Digital information has become so entrenched in all aspects of our lives and society that the recent growth in information production appears unstoppable. Figure 1.1 shows a chronological list of some of the key technological milestones that enabled the rapid and unstoppable growth of digital information production today. It is estimated that each day on Earth we generate 500 million tweets, 294 billion emails, 4 million gigabytes of Facebook data, 65 billion

WhatsApp messages, and 720,000 hours of new content added daily on YouTube[2]. In 2018, the total amount of data created, captured, copied, and consumed in the world was 33 zettabytes (ZB) – the equivalent of 33 trillion gigabytes[3]. This grew to 59 ZB in 2020 and is predicted to reach a mind-boggling 175 ZB by 2025. One zettabyte is 8,000,000,000,000,000,000,000 bits.

To help visualise these numbers, let's imagine that each bit is a £1 coin, which is around 3mm (0.1 inches) thick. One ZB made up of a stack of coins would be 2,550 light-years away. This can get you to the nearest star system, Alpha Centauri, 600 times.

Currently, each year we produce 59 times that amount of data, and the estimated compound growth rate is around 30%. Most digital information is stored in three types of locations. First is the global collection of what are called endpoints, which include all Internet of Things devices, PCs, smart phones, and all other information storage devices. Second is the edge, which includes infrastructure such as cell towers, institutional servers, and offices such as universities, government offices, banks, and factories. Third, most of the data is stored in what's known as the core – traditional data servers and cloud data centres. There are around 600 hyperscale data centres – ones with over 5,000 servers – in the world[4]. Around 39% of them are in the US, while China, Japan, the UK, Germany, and Australia account for about 30% of the total. The two largest data centres in the world are the China Telecom Data Centre, in

Hohhot, China, which occupies 10.7 million square feet, and The Citadel in Tahoe Reno, Nevada, which occupies 7.2 million square feet and uses 815 megawatts of power. To meet the ever-growing demand for digital data storage, around 100 new hyper scale data centres are built every two years.

A recent study examined the physics of information creation and determined that, assuming the current growth trends in digital content continue, the world will reach a singularity point in terms of the maximum digital information created and the power requirements to sustain it, called the Information Catastrophe[5][6].

Taking the IBM estimate that the present rate of digital content production is about 2.5 quintillion digital data bytes produced every day on Earth (2.5×10^{18} bytes, or 2.5 billion Gb)[7], and accounting that 1 byte is made up of 8 bits of digital information, the total number of bits produced on the planet daily is 2×10^{19}. From this we can easily estimate the current annual rate of digital bit production on Earth to be staggering, $N_b = 7.3 \times 10^{21}$ bits (please note this could be underestimated, as we previously quoted 33 Zb). Let us assume that f% is the annual growth factor of digital content creation on Earth. This allows the estimation of the total number of bits of information accumulated on the planet after n years of f% growth as follows:

$$N_{bits}(n) = \frac{N_b}{f} \cdot \left((f+1)^{n+1} - 1 \right)$$

(1.1)

The current estimated f% growth appears to be double digits. For instance, examining just the Bitcoin blockchain, a single block is around 1-2 Mb in size, while the entire Bitcoin blockchain is today 459.16 Gb (26 Feb. 2023)[8]. Moreover, its daily growth rate is around 0.04%, and its annual growth rate is ~16.91%.

Accounting for the fact that out-of-date digital content gets erased all the time, let us assume a conservative annual growth of digital content creation of 1%, i.e., $f = 0.01$. We need to stress that a 1% growth rate is badly underestimated, as the real rate is around 30%. However, at a 1% growth rate, we will have $\sim 10^{50}$ bits of information ~ 6000 years from now. This number of bits is very significant because it represents the approximate number of atoms on Earth. The size of an atom is $\sim 10^{-10}$ m, while the linear size of a bit of information today is 25×10^{-9} m, corresponding to about 25 nm^2 area per bit at data storage densities exceeding 1 Tb/in^2. Even assuming that future technological progress brings the bit size down to sizes closer to the atom itself, this volume of digital information will take up more than the size of the planet, leading to what we define as The Information Catastrophe[5].

If we assume more realistic growth rates of 5%, 20%, and 50%, the total number of bits created will equal the total number of atoms on Earth after ~ 1200 years, ~ 340 years, and ~ 150 years, respectively. It is important to consider that the growth of digital information today is closely linked to other factors, including population growth and increased access to information technologies in developing countries. If any of these other factors are reversed or saturated, the total number of bits of information accumulated on the planet could display saturation at some point in the future, rather than following equation (1.1).

Nevertheless, considering that our digital era started about 70 years ago and that the Earth is about 4.5 billion years old and that within a few hundred years we are looking at more digital bits being produced than all the atoms that make up the entire planet, one could say that we are literally changing the planet bit by bit. In this context, we could envisage a future world mostly computer-simulated and dominated by digital bits and computer code.

In fact, these numbers are so compelling that this is the first hint that perhaps we are already living in a simulated universe.

1.2 Eddington number

The total number of particles in the universe, protons more precisely, is known as the Eddington number, N_{Edd}, and it was first calculated in 1938[9]. Eddington estimated that the universe contains $N_{Edd} = 136 \times 2^{256}$, or about 1.57×10^{79} protons.

The objective is to estimate just how many particles are there in the visible universe and then to calculate how many years of current digital data growth it would take to equalise the number of digital bits to the number of particles in the universe.

Here we provide a more detailed analytical formula for the total number of baryons in the universe, which allows a more detailed estimation of the number of particles in the universe. Using the Friedmann equation[10] for a homogeneous, isotropic, and flat universe, we determine the well-known critical density, ρ_c, of the universe as:

$$\rho_c = \frac{3H^2}{8\pi G} \tag{1.2}$$

where $H = 2.2 \times 10^{-18}$ $1/s$ is the Hubble parameter today and $G = 6.674 \times 10^{-11}$ $m^3/Kg \times s^2$ is the gravitational constant. The critical density is then $\rho_c = 9.6 \times 10^{-27}$ Kg/m^3. The fraction of the energy density stored in baryons is $\Omega_b = 0.0485$[11], which allows us to estimate the baryons density as $\rho_b = \Omega_b \times \rho_c$. The radius of the observable universe is estimated to be about 46.5 billion light-years[12], or $L_u = 44 \times 10^{25}$ m. The volume of the observable universe is then $V_u = 4/3 \times \pi \times L^3 = 4 \times 10^{80}$ m^3. The total mass of ordinary matter in the universe, or the mass of all baryons in the observable universe, M_b, can be calculated as:

$$M_b = \rho_b \cdot V_u = \frac{H^2 \Omega_b L_u^3}{2G} \tag{1.3}$$

In order to calculate the accurate number of particles in the observable universe, we use the percentage weight abundance of elements in the universe, which are about 75% hydrogen, 23% helium and 2% heavy elements ($A > 4$) [ref. [13] and Appendix A]. Each hydrogen atom contains one proton (p^+), and one electron (e^-). Each helium atom contains $2p^+$, $2e^-$ and two neutrons (n^0). These are summarised below:

> *Hydrogen $\rightarrow$ $1p^+ + 1e^-$*
> *Helium $\rightarrow$ $2p^+ + 2e^- + 2n^0$*
> *Heavy elements $\rightarrow$ $xp^+ + xe^- + yn^0$, where x, y > 2.*

The x and y values must be known to perform an exact calculation. Since the heavy elements represent only 2% of the observable universe, one option is to use average x and y values representing all the heavy elements. However, in order to get an accurate picture, we will perform the exact calculation for all the known elements contained in the universe, including the huge range of heavy elements. The data table in Appendix A shows the percentage weight abundance of all the elements in the observable universe. The key aspect here is to realise that the percentage weight abundance must be converted into the percentage number of atoms, if one needs to estimate the total number of elementary particles in the universe and their information content. This is done using the molar mass values (units of moles/g) for each element to work out the moles (number of atoms) of each kind. This calculation shows that the percentage weight abundance of elements in the universe is very different from the percentage abundance of atoms. For example, the % weight values for hydrogen and helium are 75% and 23%, respectively, while their % numbers of atoms are 92.68% and 7.15%, respectively (see Appendix A for a full list of values).

Taking into account the internal structure of the atoms by using the exact number of p^+, e^- and n^0 in each kind of atom and using the % abundance of atoms determined, we can work out the total number of p^+, e^- and n^0 corresponding to each kind of atom. Summing up all p^+, e^- and n^0 we obtain: 108.27 electrons, 108.27 protons, and 15.6 neutrons, where the fractional values reflect the probabilistic nature of the estimates. The total statistical number of atomic constituents representative of the matter in the observable universe is then a total of 232.14 particles (p^+, e^- and n^0). This allows us to estimate the probabilities of observing p^+, e^- and n^0 in the observable universe as follows: $P_{p+} = P_{e-} = 108.27 / 232.14 = 0.466$, and $P_{n0} = 15.6 / 232.14 = 0.067$.

Considering the probabilities of each p^+, e^- and n^0, $\{P_{p+}, P_{e-}, P_{n0}\} = \{0.466, 0.466, 0.067\}$, the effective mass of a baryon is defined as:

$$m^b{}_{eff} = P_{e-}m_{e-} + P_{p+}m_{p+} + P_{n0}m_{n0} \qquad (1.4)$$

The $m_{e-} = 9.109 \times 10^{-31}$ kg, $m_{p+} = 1.672 \times 10^{-27}$ kg and $m_{n0} = 1.675 \times 10^{-27}$ kg are the rest masses of the electron, proton, and neutron, respectively. The number of baryons in the universe is then obtained by dividing the mass of all the baryons in the universe by their effective mass:

$$N_b = \frac{H^2 \Omega_b L_u{}^3}{2G \cdot m^b{}_{eff}} \qquad (1.5)$$

Performing the numerical calculations, we obtain $N_b = 1.93 \times 10^{80}$ baryons in the universe (more precisely, the total number of particles in the universe as it includes electrons, which are not baryons). This can be used to determine the exact number of electrons (N_{e-}), protons (N_{p+}) and neutrons (N_{no}), as follows:

$$N_{e-} = P_{e-}N_b \; ; N_{p+} = P_{p+}N_b \; ; N_{n0} = P_{n0}N_b \qquad (1.6)$$

The numerical values are $N_{e-} = N_{p+} = 9.027 \times 10^{79}$, and $N_{no} = 1.298 \times 10^{79}$. Since each neutron and proton contains three quarks, the total number of particles in the observable universe is: $N_{tot} = N_{e-} + 3(N_{p+} + N_{no})$, or:

$$N_{tot} = \frac{H^2 \Omega_b L_u^{\ 3}}{2G \cdot m^b_{eff}} \left(P_{e-} + 3\left(P_{p+} + P_{n0} \right) \right) \tag{1.7}$$

The calculations indicate that the observable universe contains $N_{tot} = 4 \times 10^{80}$ particles[14], which is in good agreement with Eddington's number, which gives only the number of protons.

Taking our current annual growth rate in digital data production as 30% and using (1.1), one can calculate that ~ 240 years from now, the number of digital bits will equal all the particles in the universe. Again, these numbers are so staggering, and the time scales are equally minuscule relative to cosmic time, that one would naturally wonder whether the whole universe is not already some kind of digital construct. In the next chapter, we briefly examine some physical facts that could be interpreted as evidence of a simulated universe.

2

Physical Facts Supporting the Simulated Universe Hypothesis

The concept that reality is an illusion is not new. The earliest records of this philosophy go as far back as ancient Greece, when the question *"What is the nature of our reality?"* gave birth to two principal viewpoints or ideologies: materialism and idealism.

Materialist thinkers like Democritus (460 BC) considered matter to exist objectively independent of the mind and consciousness. Materialists view consciousness as a special property of matter, and mind cannot exist without matter, while matter can exist without mind. These views align with the mainstream scientific consensus today. Matter existed before the appearance of thinking biological entities like humans.

Idealist ancient thinkers like Plato (427 BC) and Aristotle (384 BC) considered mind and spirit as the abiding reality and matter just a manifestation or illusion. Idealist thinkers view the mind or spirit as emanating from the divine and being of the same nature as the divine. Religions around the world are strongly rooted in idealistic philosophy, which leaves open the possibility of supernatural existence, power, and interference.

Fast-forward to modern times, and idealism has morphed into a new philosophy implying that both the material world and consciousness are part of a simulated reality. The concept of a simulated universe emerged as a modern extension of idealism, driven by recent technological advancements in computing and digital technologies. Just as Plato proposed that the material world is a mere reflection of higher forms, the simulation hypothesis suggests that our reality might be a digital projection. In both cases, the true nature of reality transcends the physical.

Technological progress has played a pivotal role in the morphing of idealism into the simulation hypothesis. The exponential growth of computing power and the development of immersive digital environments have fuelled speculation about the plausibility of simulating entire universes. As humans create increasingly sophisticated simulations, the line between the virtual and the real becomes blurred, raising questions about the limits of our ability to distinguish between the two.

Moreover, the simulation hypothesis aligns with the observation that technological civilizations might eventually possess the capability to create simulated universes. This notion echoes Plato's idea of a higher realm creating the imperfect material world. Thus, the convergence of technological advancements and philosophical contemplation has led to the simulation hypothesis as a contemporary iteration of idealism.

The origins of this iteration can be traced back to the philosopher Jean Baudrillard, who introduced the concept of "simulacra"[15]. Baudrillard postulated that simulations are imitations of real-world processes, culminating in what he termed as "simulacra" – the ultimate detachment from reality. One of the earliest propositions within the scientific community came from Brian Whitworth. He dared to speculate that the physical world, as we perceive it, could potentially be a manifestation of virtual reality[16]. However, it was the emergence of the simulation hypothesis itself that truly captured the imagination of many. Hans Moravec, a researcher in robotics and artificial intelligence, presented an early version of this hypothesis[17]. Moravec's ideas hinted at the possibility that our reality could be a construct meticulously designed by an advanced civilization, reflecting their technological prowess.

It was the philosopher Nick Bostrom who brought the simulation hypothesis into sharper focus in 2003[18]. Bostrom's seminal work postulated a compelling argument: an advanced civilization could eventually create simulations so advanced that they are indistinguishable from reality. Bostrom's vision encompassed a scenario where participants within these simulations remain oblivious to their fabricated existence. This groundbreaking proposition ignited debates and discussions across academic circles, urging scholars to grapple with the intricate implications of living within a simulation.

Taking this hypothesis to a new dimension, physicist Seth Lloyd proposed an astonishing concept. He proposed that the universe itself might function as an elaborate quantum computer, computing its own existence[19].

In 2016, the simulation hypothesis received a resounding endorsement from an unexpected source: Elon Musk. The tech visionary postulated that as technology advances, the boundary between simulated experiences and authentic reality will blur to the point of being indistinguishable. Musk's bold statement, "We're most likely in a simulation," reverberated across popular culture and invigorated further contemplation[20].

If our physical reality is a simulated construct rather than an objective world that exists independently of the observer, then what empirical evidence do we have to support this? In particular, we are interested in exploring the concept not from a philosophical angle but instead using fundamental physics principles and empirical data. This is the prime objective of this book.

2.1 Pixalation of the universe

At the heart of the pixelated reality hypothesis is the idea that, at the most fundamental level, space, time, and energy are not continuous but rather possess discrete characteristics, much like the pixels in a digital image. They are fundamentally quantized, meaning they can only take on discrete values. This notion finds its roots in the Planck scale, where space, time, and energy are thought to have their smallest, indivisible units: the Planck length, Planck time, and Planck energy. These Planck scales represent the ultimate granularity of the energy and space-time fabric, and are a cornerstone of modern physics.

Quantum mechanics provides a wealth of evidence supporting the pixelated nature of our reality. One key piece of evidence comes from the quantization of energy levels in atomic systems. Electrons in atoms can only occupy certain energy levels with discrete transitions between them, just like the steps on a staircase. This quantization of energy strongly resonates with the idea of a pixelated reality, where energy levels are confined to distinct values. The Planck units, acting as fundamental constants of nature, play a crucial role in reinforcing the idea of pixelation. These units provide a natural framework for understanding the smallest possible intervals of length, time, and energy. The fact that these units define the scale at which classical physics breaks down suggests an inherent granularity to our reality.

If our reality were indeed a simulation, it would be logical for the simulation to employ discrete units, mirroring the design principles used in creating virtual worlds. The pixelation aligns with the concept of information processing, where the simulation is structured based on a grid of discrete units of space, time, and energy. To put it simply, our world is inherently pixelated, much like a virtual reality world, and the notion that our reality is simulated gains traction when considering the pixelated nature of our world.

2.2 The universe is mathematical

The simulated universe theory suggests that the laws of physics and the entire fabric of our reality can be understood as lines of code within a vast computational system. This idea gains credibility from the remarkable correspondence between mathematical equations and the behaviour of the physical world. Throughout history, scientists have discovered equations that elegantly describe natural phenomena, from the motion of planets to the behaviour of subatomic particles. The ubiquity of mathematics in explaining physical processes hints at a deeper connection between mathematics and the fundamental structure of the universe. The prevalence of mathematical equations, numbers, and symmetries throughout the natural world suggests that our reality could indeed be the product of an intricate computational simulation.

Numbers and geometry are not mere abstract concepts but appear to be the fundamental language of the cosmos. From the Fibonacci sequence found in the arrangement of leaves on plants to the golden ratio observed in seashells, nature exhibits patterns that can be described mathematically. One possibility is that these mathematical patterns are not coincidental but are instead inherent to the underlying computational framework that generates our reality.

Geometrical symmetries, such as fractals, tessellations, and the Platonic solids, are prevalent throughout the natural world. These symmetries mirror the symmetrical patterns that often arise in computational simulations. The presence of such symmetries across different scales, from galaxies to subatomic particles, supports the idea that our universe is governed by a consistent computational mechanism that manifests as mathematical patterns.

One of the most compelling arguments for the simulated universe theory is the role of symmetries in shaping the laws of physics and the behaviour of matter and energy. Symmetry principles, such as conservation laws and gauge symmetries, underpin our understanding of fundamental forces and particles. For instance, Noether's theorem establishes a profound link between symmetries and conservation laws. This relationship between symmetries and fundamental physical principles suggests that the universe operates according to a structured computational code that enforces these symmetrical patterns. Furthermore, the Standard Model of Particle Physics, which describes the fundamental particles and forces in the universe, is deeply rooted in mathematical symmetries.

2.3 Speed of light

One of the empirical curiosities in the realm of physics that seemingly supports the simulation hypothesis is the maximum speed limit in our universe - the speed of light. At the heart of Einstein's theory of relativity lies the postulate that the speed of light in a vacuum is an absolute constant, denoted as c, with a value of approximately 299,792,458 metres per second. This speed serves as a cosmic speed limit, beyond which nothing can travel. This unique feature of the universe has profound implications for our understanding of space and time. It leads to the famous time dilation effect, where an object approaching the speed of light experiences a slowing of time as observed by an external observer.

In the realm of virtual reality, a similar concept exists: the processing power limit. In a virtual environment, the speed at which the computer can process information and render the simulation defines the maximum achievable performance. Just as the speed of light sets the maximum speed achievable in our universe, the processing power of the computer defines the limits of the virtual world. This parallel between the two concepts opens up a thought-provoking avenue when considering the simulation hypothesis.

A noteworthy comparison between the simulated universe theory and the speed of light is the phenomenon of relativistic time dilation. According to Einstein's theory, as an object with mass approaches the speed of light, time appears to slow down for that object as observed by a stationary observer. This phenomenon can be likened to a processor overload in a virtual reality system. When a computer is overloaded with processing tasks, it struggles to keep up with the demands of the simulation, leading to a noticeable slowdown in performance.

Similarly, in regions of high mass concentration, such as around black holes, space-time becomes highly curved, leading to significant gravitational effects. This curvature can be considered analogous to a "processor overload" within the simulation, causing time to slow down for objects in the vicinity.

This intriguing parallel raises questions about whether the observed relativistic effects are a consequence of a simulated universe trying to manage computational demands, analogous to a computer's struggle with processing overload.

2.4 Quantum entanglement

Perhaps the most supportive evidence for the simulation hypothesis comes from quantum mechanics. Quantum entanglement is a phenomenon that occurs when two or more particles become correlated in such a way that the state of one particle is intrinsically linked to the state of the other(s), regardless of the distance that separates them. This correlation appears to defy our classical intuitions, as it suggests that changes in the state of one

17

particle instantaneously affect the state of the entangled particle, seemingly transcending the constraints of space and time. Einstein, Podolsky, and Rosen famously characterised this concept in their EPR paradox paper in 1935[21]. They proposed that the apparent non-local interactions implied by quantum entanglement were inconsistent with the principles of local realism, prompting what Einstein famously referred to as "spooky action at a distance".

The non-local nature of quantum entanglement challenges our conventional understanding of space and locality. In a classical sense, interactions between objects are limited by the speed of light, imposing a universal speed limit on information transfer. However, quantum entanglement seems to defy this limit, suggesting that particles can instantaneously influence each other's states, regardless of their spatial separation.

In the context of the simulation hypothesis, quantum entanglement takes on a new layer of significance. In a simulated universe, the laws of physics and the behaviour of particles are governed by the underlying code of the simulation. This framework suggests that the concept of distance as we understand it in our physical reality might not be a fundamental aspect of the simulated universe. According to this line of reasoning, the phenomenon of non-locality observed in quantum entanglement could be explained by considering that, within the confines of a simulated reality, all points are equidistant from the central processing unit, or "processor". In such a scenario, interactions between particles would not be restricted by spatial separation because space itself is not inherently real. Instead, space is a construct generated by the simulation. This aligns with the idea that the simulation is capable of manipulating the properties of space and distance to facilitate instantaneous interactions between entangled particles.

The scientific community continues to explore alternative explanations for quantum entanglement and non-locality that do not necessarily require a simulated universe. However, the connection between quantum entanglement, non-locality, and the simulation hypothesis is very plausible, and it raises profound questions about the nature of reality, challenging our understanding of space, time, and causality.

2.5 Wave-particle duality

One of the most enigmatic and counterintuitive phenomena that challenges our understanding of reality is the wave-particle duality exhibited by elementary particles in the quantum realm. Wave-particle duality is a cornerstone of quantum mechanics, the branch of physics that deals with the behaviour of matter and energy at atomic and subatomic scales. It states that elementary particles can exhibit behaviours characteristic of both waves and particles, depending on how they are observed or measured. For instance, an electron can behave like a wave, displaying interference patterns in certain experiments, but can also behave as a discrete particle, producing well-defined impact patterns in other experiments. This dual nature of particles baffled early physicists and led to the development of the famous double-slit experiment, where particles are sent through two slits and create an interference pattern on a screen, suggesting their wave-like nature even when sent through one particle at a time. This phenomenon suggests that particles such as electrons and photons possess both wave-like and particle-like properties, and these properties seem to manifest only when the wave function collapses upon observation or measurement.

This intriguing behaviour finds a curious resonance with the simulated universe hypothesis, which proposes that our reality might be a complex simulation akin to a virtual reality, where matter and phenomena are observer-dependent, existing only when observed. The link between wave-particle duality and the simulated universe hypothesis is not direct, but it opens up intriguing avenues for exploration.

In the simulated universe scenario, reality could indeed be observer-dependent, similar to the way particles exhibit properties upon observation. The act of observation or measurement in quantum mechanics collapses the wave function, determining the particle's state. Similarly, if our universe is a simulation, the rendering of certain aspects of reality could be determined by the observer's interaction, much like particles becoming "real" upon measurement.

No question – No answer
No observation – No measurement

In a simulated universe, energy optimisation would be necessary, so resources might be allocated only when needed, analogous to particles exhibiting properties only when measured. This mirrors the quantum concept of particles "collapsing" into a specific state upon observation. In a simulation, rendering all aspects of reality all the time would be resource-intensive, so rendering reality on-demand based on observation could optimise energy usage.

A final observation is the fact that quantum mechanics introduces inherent uncertainty and probabilistic behaviour at the quantum level. This aligns with the notion of a simulated universe where the behaviour of particles and events could be probabilistic in nature, which is similar to the way a computer simulation can use randomness to generate diverse outcomes.

2.6 The Big Bang and the inflation theory

The big bang theory stands as the prevailing explanation for the origin of the universe, offering remarkable insights into its early moments and subsequent evolution. However, this theory is not without its share of enigmatic aspects and apparent contradictions in the realm of physics. The big bang theory, proposing that the universe originated from an infinitesimally dense and hot state approximately 13.8 billion years ago, has garnered widespread acceptance within the scientific community. While it successfully describes the expansion of the universe, the cosmic microwave background radiation, and the formation of galaxies, it leaves certain questions unanswered and introduces perplexing contradictions in the field of physics.

For instance, there are no clear answers to what exactly banged and what triggered the Big Bang in the first place.

The very concept of the initial singularity, a point of infinite density and temperature, challenges the limits of our current understanding of physics. The laws of physics break down in such conditions, rendering our ability to describe the events at that moment severely limited.

The observed nearly flat geometry of the universe contradicts the predictions of the big bang theory without the inclusion of a mysterious dark energy.

The inherent flatness of the universe's geometry also implies a fine-tuning conundrum.

The big bang theory doesn't provide insights into the nature of dark matter and dark energy, which collectively constitute about 95% of the universe, while the remaining 5% represents what we perceive as the visible universe today[22].

The overwhelming preponderance of matter over antimatter in the universe, known as baryon asymmetry, lacks a convincing explanation within the big bang framework.

The early cosmic inflationary period after the big bang appears to violate the special relativity postulate that nothing can travel faster than the speed of light. According to the theory, during the inflationary epoch, which started about 10^{-32} seconds after the big bang, the universe expanded exponentially in volume by a factor of at least 10^{78}. This is equivalent to an expansion in linear dimensions by a factor of at least 10^{26} in each of the three dimensions. Although it is not known exactly when the inflationary epoch ended, it lasted a tiny fraction of a second, so one would get expansion velocities far exceeding the speed of light. While inflation theory elegantly explains the uniformity of the cosmic microwave background radiation, its driving mechanism and the precise physics governing inflation remain theoretical.

Understanding these enigmatic components is vital to comprehending the universe's evolution, and the simulated universe hypothesis offers an intriguing alternative perspective, proposing that our universe is an intricate construct within an advanced simulation and the big bang is just the point of start of the simulation itself.

This hypothesis motivates the development of novel methodologies to discern between a simulated universe and a naturally evolving one, potentially leading to new discoveries.

2.7 Pauli's exclusion principle

Electronic states in atoms are fully described by four principal quantum numbers: a) the principal quantum number, n; b) the orbital angular momentum quantum number, ℓ; c) the magnetic quantum number, m_ℓ; d) the secondary spin quantum number, m_s.

The electrons occupy atomic shells according to Pauli's exclusion principle[23], which states that two or more identical fermions cannot simultaneously occupy the same quantum state within a quantum system. This principle underpins the stability of matter and is pivotal to our understanding of atoms, chemical bonding, the diversity of chemicals, forms of matter, life, and various other processes and phenomena in the universe.

In the case of electrons in atoms, this means that it is impossible for two electrons in a multi-electron atom to have the same values of the four quantum numbers described above. For example, if two electrons reside in the same orbital, then their n, ℓ, and m_ℓ values are the same, so their m_s must be different, imposing that the electrons must have opposite half-integer spin projections of $1/2$ and $-1/2$.

Remarkably, in its insistence on the distinguishability of particles, this principle bears an uncanny resemblance to the rules of coding and programming, specifically in the context of defining variables that must be distinguishable.

In programming, ensuring the distinction of variables is paramount to maintaining order and predictability in the execution of code. Similarly, the exclusion principle mandates that fermions (particles with half-integer spin) maintain their individuality even in the most densely populated regions of space. This apparent congruence prompts us to question whether there might be an intrinsic link between the nature of the universe and the principles governing computational simulations.

This raises thought-provoking questions:

1) *Does the apparent correlation between the exclusion principle and coding practices imply that our universe is intricately constructed according to a logical framework similar to computer programming?*

2) *Is the Pauli's exclusion principle a manifestation of the simulation's underlying code?*

23

2.8 Cambrian explosion. Where did all the information come from?

The Cambrian Explosion stands as a pivotal event in the history of life on Earth, representing a period of remarkable biological innovation and diversification that unfolded approximately 541 million years ago during the Cambrian period and lasted for around 40-50 million years. This phenomenon is characterised by the sudden appearance of a wide array of complex and diverse multicellular organisms in the fossil record, marking a stark departure from the relatively simpler forms of life that had dominated the preceding eras. Prior to the Cambrian Explosion, life on Earth primarily consisted of single-celled microorganisms and simple multicellular organisms. This sudden emergence of new forms of biological life is associated with a burst of genetic information, and, for this reason, the Cambrian Explosion is also known as the Cambrian Information Explosion.

The Cambrian Information Explosion has sparked extensive scientific inquiry into its causes. One prominent explanation is the designer hypothesis[24], which suggests that an intelligent agent or external force played a role in orchestrating the rapid appearance of diverse and complex life forms during the Cambrian period. Proponents of this hypothesis argue that the complexity and diversity seen in the Cambrian fossil record are beyond what could be achieved through natural evolutionary processes alone. They propose that an intelligent designer could have been responsible for imparting the genetic information necessary to bring about these intricate biological structures.

However, the designer hypothesis has been met with substantial criticism from the scientific community. Detractors argue that invoking an intelligent designer raises more questions than it answers, as it sidesteps the need for a comprehensive and evidence-based explanation rooted in natural processes.

The majority of researchers within the field of evolutionary biology advocate for naturalistic explanations of the Cambrian Information Explosion. They propose that a combination of factors, including environmental changes, the evolution of new genetic regulatory networks, and the interplay between developmental processes, could have contributed to the rapid diversification of life during the Cambrian period.

At the heart of the statistical challenge lies the complexity of forming even a single functional protein through genetic mutations. Proteins are the workhorses of life, carrying out a vast array of biological functions, and they are built from sequences of amino acids encoded in the DNA. The sequence of amino acids determines a protein's three-dimensional structure and, consequently, its function. The probability of a random mutation leading to a specific functional protein sequence is incredibly low due to the astronomical number of possible amino acid combinations. Some proponents of the designer hypothesis point to these calculations to argue that the odds of such fortuitous events occurring within the timeframe of the Cambrian Explosion are outright impossible. For instance, the probability of attaining the correct sequencing at random even for a single modest protein (150 or so amino acids in length) was calculated to be less than 1 chance in 10^{65} [25]. If a mutation took place every second, it would take longer than the age of the universe ($\sim 5\times10^{17}$s) before a single protein would be randomly formed.

These statistical challenges raise questions about the plausibility of purely naturalistic mechanisms accounting for the rapid diversification of complex life during the Cambrian period. Some proponents of the designer hypothesis contend that an external intelligent agent or designer could have played a role in guiding these intricate processes, ensuring that the genetic information necessary for the emergence of diverse and functional organisms was present.

The simulation hypothesis is fully aligned with the intelligent design theory, and it offers an alternative explanation to the Cambrian Information Explosion, which could be simply explained as an external input of information (genetic information) / intervention directly into the code.

3

Brief Introduction to Shannon's Information Entropy

Shannon gave the mathematical formulation of the amount of information extracted from observing the occurrence of an event in his 1948 seminal paper[26]. Ignoring any particular features of the event, the observer, or the observation method, we would like to develop a usable measure of the information we get from observing an event with a probability p of occurrence. By event, we mean any process or observation whose probability of occurrence is p. Such an event could be the observance of a symbol, the tossing of a coin, the sequencing of a genome, and so on, and we want to determine the information extracted I, in terms of the probability p.

Shannon developed his theory using an axiomatic approach:

1) He defined information (I) extracted from observing an event as a function of the probability (p) of the event to occur or not, $I(p)$.

2) The information measure is a continuous positive function of the probability, $I(p) \geq 0$.

3) An event that is certain, i.e. $p = 1$, gives therefore no information from its occurrence, so $I(1) = 0$.

4) Assuming that for N independent events of individual probabilities p_j the joint probability p is the product of their individual probabilities, then the information we get from observing the set of N events is the sum of the individual event's information, $I(p) = I(p_1 p_2 \cdots p_N) = I(p_1) + I(p_2) + \cdots + I(p_N)$.

Shannon identified that the only function satisfying these axiomatic properties is a logarithmic function, and, for an event whose probability of occurring is p, the information extracted from observing the event is:

$$I(p) = -\log_b p = \log_b(1/p) \qquad (3.1)$$

where b is an arbitrary base positive integer, which gives the units of information. Units of bits are obtained when $b = 2$, trits when $b = 3$, and nats when $b = e$, i.e., Euler's number. The natural choice of $b = 2$, resulting in bits, is dictated by the current digital technologies, making this a convenient choice. For an arbitrary choice of the base "b", the information function can be returned in different units using the logarithm base change formula:

$$I(p) = \log_b(1/p) = \frac{\log_a(1/p)}{\log_a b} \qquad (3.2)$$

For example, if we want to convert information expressed in nuts into bits, then $b = 2$, $a = e$, and $I(p) = \ln(1/p)(nuts) = \ln(2) \cdot \log_2(1/p)(bits)$.

In order to derive Shannon's information entropy, it is useful to first introduce the following parameters:

$N =$ *the total number of single characters x in a given set:* $x_1, x_2, \ldots, x_N$;

$U =$ *total number of distinct single characters in a set,* $1 \le U \le N$;

$X =$ *the set of unique characters,* $X = x_1, x_2, \ldots, x_U$

$m =$ *the m-block size, or the number of single characters combined to form a new character, $1 < m < N$;*

$ss =$ *the fixed number of characters of sliding m-block from left to right, $1 < ss < m$;*

$\Omega =$ *the number of distinct m-blocks in a set;*

$IE =$ *information entropy;*

Let us assume that a set contains N characters, and out of the total of N characters, the set X contains U distinct characters $X = \{x_1, x_2, \ldots, x_U\}$, where $U \leq N$ and a probability distribution $P = \{p_1, p_2, \ldots, p_U\}$ is defined on X, such that each unique character x_j has a probability of occurring $p_j = p(x_j)$, where $p_j \geq 0$ and $\sum_{j=1}^{U} p_j = 1$.

We want to estimate the amount of information one gets from each character in the set. If we observe the symbol x_j, we will get $log(1/p_j)$ information from that particular observation. Over n observations, we get approximately $n \times p_j$ occurrences of the symbol x_j. Thus, for n independent observations, the total information I extracted is:

$$I = \sum_{j=1}^{U} (p_j \cdot n) \cdot \log_b \frac{1}{p_j} \tag{3.3}$$

From (3.3), the average information we get per character is obtained by dividing both sides of the relation by n:

$$\frac{I}{n} = \frac{1}{n} \sum_{j=1}^{U} (p_j \cdot n) \cdot \log_b \frac{1}{p_j} = \sum_{j=1}^{U} p_j \cdot \log_b \frac{1}{p_j} \tag{3.4}$$

According to Shannon, the average information extracted per character, or the number of bits of information per character for the set X is:

$$I = IE = -\sum_{j=1}^{U} p_j \cdot \log_b p_j \tag{3.5}$$

Since this relation resembles Boltzmann thermodynamic entropy, the average information content per event is known as Shannon's Information Entropy (*IE*). For the entire set, the total bit content of the set is $N \times IE$.

Since $\lim_{k \to 0} k \cdot \log(1/k) \to 0$, then we can assume $p_j \cdot \log_b p_j = 0$, when $p_j = 0$. Showing that information extracted from observing an event is zero if the event is certain or if the event has a zero probability of occurring.

For a continuous probability distribution, *p(x)*, Shannon's information entropy is:

$$IE = -\int p(x) \cdot \log_b p(x) \cdot dx \qquad (3.6)$$

For a given set of characters/events, the maximum *IE* is obtained when the set contains U distinct events x_j that have equal probabilities of occurring, $p_j = 1/U$, so:

$$IE_{max} = \log_b U \qquad (3.7)$$

The *IE* (3.5) was computed when the probabilities of occurrence refer to single characters within a set. However, a useful extrapolation could be the generalisation of relation (3.5) to m-block *IEs*[27]:

$$IE^{(m)} = -\sum_{j=1}^{\Omega} p_j^{(m)} \cdot \log_b p_j^{(m)} \qquad (3.8)$$

where instead of single characters, combinations of *m* characters are used to define a new set of characters called m-blocks. For instance, m-blocks could be whole words in a message instead of single characters. In this case, $p^{(m)}$ is the probability of the m-block characters to occur and the summation extends over all possible combinations of distinct m-blocks. The maximum number of distinct m-blocks of the new set of m-block characters constructed from a given set of N single characters, containing U distinct characters, is:

$$\Omega = U^m \tag{3.9}$$

The maximum value of the IE theoretically permitted for the new set of m-blocks is:

$$IE^m{}_{max} = \log_b \Omega = \log_b U^m \tag{3.10}$$

Combining (3.7) and (3.10) we deduce that: $IE^m{}_{max} = mIE_{max}$, so using m-blocks increases the IE value by a factor of m relative to the set of single characters.

In order to clarify the methodology proposed in this invention, it is useful to show a few examples. Let's assume that a given set of characters contains only two distinct single characters, so $U = 2$. Using binary units, $b = 2$, and relation (3.9), then:

If $m = 1$, $\Omega = U^m = 2$, indicating that we have two possible states and each state encodes $IE = 1$ bit per character:

{0, 1}

If $m = 2$, $\Omega = U^m = 4$, indicating that we have four possible states and each state encodes $IE = 2$ bits per character:

{01, 00, 10, 11}

If $m = 3$, $\Omega = U^m = 8$, indicating that we have eight possible states and each state encodes $IE = 3$ bits per character:

{000, 001, 011, 101, 110, 100, 111, 010}

If $m = 4$, $\Omega = U^m = 16$, indicating that we have 16 possible states and each state encodes $IE = 4$ bits per character:

{0000, 0001, 0011, 0111, 1111, 1001, 0100, 0101, 0111, 0010, 1100, 1110, 1101, 1110, 1000, 0110}

If the set has N characters and we take $m = N$, then $\Omega = U^m = 2^N$ is the number of possible states, and each state encodes $IE = N$ bits per character.

3.1 Generating a set of m-blocks

A new set constructed using m-blocks from a given set of N single characters, containing U distinct characters, will have $\Omega = U^m$ possible distinct m-blocks / characters. We are showing a method for generating a set of m-blocks from a set of single characters and working out how many m-block elements will be in the newly formed set of m-blocks. The new set of m-blocks contains N_m elements. The set of m-blocks is constructed by sliding the m-block segment from left to right in step size ss, where the condition on ss is now $1 < ss \leq m$. The newly formed set of m-block elements contains a number of characters given by:

$$N_m = \frac{N}{s} - m + s \qquad (3.11)$$

It is important to observe that the values of N, m, and ss are selected so that the sliding procedure produces a set of N_m integer elements. To clarify this procedure, let's observe few examples again. Let's assume a random set of $N = 16$ single characters and $U = 2$:

0110101101011100

Taking $ss = 1$ and $m = 2$, we generate the new set of m-block characters by sliding the m-block segment of two single characters from left to right in steps of 1. This results in the following set containing $N_m = 15$ elements, as dictated by (3.11):

{01, 11, 10, 01, 10, 01, 11, 10, 01, 10, 01, 11, 11, 10, 00}

Constructing another m-block set with $ss = 1$ and $m = 3$, according to (3.11) we obtain the following set of $N_m = 14$ elements:

{011, 110, 101, 010, 101, 011, 110, 101, 010, 101, 011, 111, 110, 100}

Using this procedure, a new set of any m-block size, with $m \leq N$, can be generated. In order to ensure that the IE per window can take all possible values between zero and the theoretical maximum value, $log_2 U^m$ the imposed condition is: $N_m \geq U^m$. Using (3.11) and solving for N, we obtain:

$$N = s \cdot U^m + m \cdot s - s^2 \qquad (3.12)$$

3.2 Calculating the IE of a digital file

Any digital file is composed of 0s and 1s in machine code / binary language. This can be seen as a set of N characters containing two distinct characters, 0 and 1, so $U = 2$, $X = \{0,1\}$, and a probability distribution $P = \{p_0, p_1\}$. Assuming one is able to decompose a digital file into its binary form, for any digital file, one can compute the IE of the file. To demonstrate this process, we use the same random set of $N = 16$ single bits shown in the previous section:

0110101101011100

If the bits within this set would occur with equal probabilities (1/2), then the set would have $IE = 1$, and a total entropy of $N \times IE$ = 16 bits of information. However, taking single bits, the above set has the following probability distribution:

$$P = \left\{ p_0 = \frac{7}{16}, p_1 = \frac{9}{16} \right\}$$

resulting in:

$$IE = -\left(\frac{7}{16} \log_2 \left(\frac{7}{16} \right) + \frac{9}{16} \log_2 \left(\frac{9}{16} \right) \right) = 0.989$$

Hence, the IE per bit of this set of bits is 0.989 bits instead of 1 bit, and the total IE of the set is 15.84 bits instead of 16 bits.

If instead of single bits we now consider m-blocks, taking $ss = 1$ and $m = 2$, the new set of m-blocks contains $\Omega = U^m = 4$ possible states, and each state encodes $E = \log_2 \Omega = 2$ bits per m-block element. The four possible states of the new set, or the four unique m-block characters are $X = \{0\ 0,\ 0\ 1,1\ 0,1\ 1\}$, having a probability distribution $P = \{p_{00},\ p_{01},\ p_{10},\ p_{11}\}$.

The new set is obtained by sliding the m-block segment of two characters from left to right in steps of 1, resulting in the following set containing $N_m = 15$ elements, as dictated by (3.11):

$$\{01,\ 11,\ 10,\ 01,\ 10,\ 01,\ 11,\ 10,\ 01,\ 10,\ 01,\ 11,\ 11,\ 10,\ 00\}$$

If the m-blocks within this new set would occur with equal probabilities $(1/4)$, then the set would have $IE = 2$, and a total entropy of $N_m \times IE = 30$ bits of information. However, counting the occurrences, the above set has the following probability distribution:

$$P = \left\{ p_{00} = \frac{1}{15}, p_{01} = \frac{5}{15}, p_{10} = \frac{5}{15}, p_{11} = \frac{4}{15} \right\}$$

resulting in:

$$IE = -\left(\frac{1}{15}\log_2\left(\frac{1}{15}\right) + \frac{5}{15}\log_2\left(\frac{5}{15}\right) + \frac{5}{15}\log_2\left(\frac{5}{15}\right) + \frac{4}{15}\log_2\left(\frac{4}{15}\right) \right) = 1.826$$

Hence, the IE per m-block character of this new set is 1.826 bits instead of 2 bits, and the total IE of the set is 27.39 bits instead of 30 bits.

3.3 Calculating the IE of a genomic sequence

A DNA / RNA sequence can be represented as a long string of the letters A, C, G, and T. These represent the four nucleotides: adenine (A), cytosine (C), guanine (G), and thymine (T) (replaced with uracil (U) in RNA sequences). All the biological information about a given living organism is fully encoded in its DNA / RNA.

Hence, within Shannon's information theory framework, a typical genome can be represented as a 4-state probabilistic system, with $U = 4$ distinctive events, X={A,C,G,T} and probabilities p ={p_A,p_C,p_G,p_T}. Assuming the events have equal probabilities of occurring, using digital information units, i.e. base of the logarithm is 2, and equation (3.5) for $U = 4$, we determine that the maximum Shannon *IE* is 2, indicating that the maximum information encoded per nucleotide is 2 bits, i.e. $X = \{A,C,G,T\}=\{00,01,10,11\}$.

Using the maximum information content per nucleotide, we convert the entire genomic sequence into a string of 0s and 1s, by replacing A with 00, C with 01, G with 10 and T with 11. In order to clarify the method, let us consider a random RNA segment consisting of $N = 20$ characters:

CAATTTTGACTTCATTTTAG

Applying the proposed conversion, the following set of 40 characters containing the two distinctive characters 0 and 1 is obtained:

0100001111111111000011111010011111111110010

The *IE* of the RNA sequence can then be computed using the same method described in the previous example for digital data. However, converting the genomic characters into binary, $X = \{A,C,G,T\}=\{00,01,10,11\}$, is not necessary, because one could calculate the *IE* of the sequence directly using the genomic characters. Let us consider a genome subset randomly generated, consisting of $N = 34$ letters:

CACTTATCATTCTGACTGCTACGGGCAATATGTG - Original subset
↓
CACTTATCATACTGACTGCTACGGGCAATATGTG - Mutated subset

If the letters within the original subset would have equal probabilities of occurring (1/4), then the subset would have $IE = 2$ bits and a total entropy of $N{\times}IE = 68$ bits of information. However, the original subset has the following probability distribution and corresponding information entropy:

$$P=\left\{p_A=\frac{8}{34},p_C=\frac{8}{34},p_G=\frac{11}{34},p_T=\frac{7}{34}\right\}$$

$$IE=-\left(\frac{8}{34}\log_2\left(\frac{8}{34}\right)+\frac{8}{34}\log_2\left(\frac{8}{34}\right)+\frac{11}{34}\log_2\left(\frac{11}{34}\right)+\frac{7}{34}\log_2\left(\frac{7}{34}\right)\right)=1.978$$

Therefore, instead of 68 bits to encode the sequence, the entropy is 67.25 bits. However, this approach is incomplete because it ignores the correlations between the individual symbols within the subset. In order to incorporate these possible correlations, the m-block information entropy must be calculated[28]. The choice of m-blocks here is based on the observation that coding sequences are via codons, i.e., blocks of 3 nucleotides. Hence, with $m = 3$, each codon will be considered a distinct symbol. In this situation, the new set of m-blocks contains a maximum of $\Omega = U^m = 4^3 = 64$ possible states, and each state encodes $IE = \log_2 \Omega = 6$ bits per m-block element. Assuming the readout could start at any nucleotide and counting each three adjacent nucleotides from left to right, so $ss = 1$, the original subset contains a set of 32 three-letter combinations, with 29 distinct combinations:

$$X = \text{\{CAC, ACT, CTT, TTA, TAT, ATC, TCA, CAT, ATT, TTC,}$$
$$\text{TCT, CTG, TGA, GAC, TGC, GCT, CTA, TAC, ACG, CGG,}$$
$$\text{GGG, GGC, GCA, CAA, AAT, ATA, ATG, TGT, GTG\}}$$

and their probability distribution:

$$P=\left\{\begin{array}{l}\frac{1}{32},\frac{2}{32},\frac{1}{32},\frac{1}{32},\frac{2}{32},\frac{1}{32},\frac{1}{32},\frac{1}{32},\frac{1}{32},\frac{1}{32},\frac{1}{32},\frac{2}{32},\frac{1}{32},\frac{1}{32},\frac{1}{32},\\[2mm]\frac{1}{32},\frac{1}{32},\frac{1}{32},\frac{1}{32},\frac{1}{32},\frac{1}{32},\frac{1}{32},\frac{1}{32},\frac{1}{32},\frac{1}{32},\frac{1}{32},\frac{1}{32},\frac{1}{32},\frac{1}{32}\end{array}\right\}$$

The maximum possible *IE* of the subset is $log_2 29 = 4.858$ bits, but according to (3.8), the actual entropy is 4.813 bits.

The reader might wonder why the calculation of *IE* for digital data and genomes is relevant to this book. This exercise is particularly important because in Section 6, we will discuss the second law of information dynamics and its application to digital data and genetic information systems. The calculation of the *IE* of a given information dataset is of prime importance to demonstrate the universal applicability of the second law of infodynamics and its implications for the simulated universe theory.

4

Information is the
Fifth State of Matter

In 1961, Landauer proposed the idea that a digital information bit is physical and has a well-defined energy associated with it[29][30]. This is known as Landauer's principle, and it was recently confirmed experimentally[31][32][33][34]. In a different study, using Shannon's information theory and thermodynamic considerations, Landauer's principle has been extended to the Mass-Energy-Information (M/E/I) equivalence principle[35]. The M/E/I principle states that information is a form of matter, it is physical, and it can be identified by a specific mass per bit while it stores information, or by an energy dissipation following the irreversible information erasure operation, as dictated by Landauer's principle. The M/E/I principle has been formulated while strictly discussing digital states of information. However, because Shannon's information theory is applicable to all forms of information systems and is not restricted only to

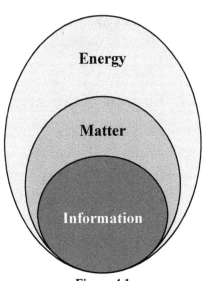

Figure 4.1

digital states, the author extrapolated the applicability of the M/E/I principle to all forms of information, proposing that information is the 5th state of matter[35][36].

These ideas, regarded as information conjectures, are truly transformational because, without violating any laws of physics, they have created a bridge between pure mathematics and physics, essentially "physicalizing" the mathematics. The concept of physicalizing mathematics has profound implications for the way we think about the whole universe because it shows that the universe is fundamentally mathematical and can be seen as emerging from information, i.e. "it from bit", a concept coined by the legendary physicist John Wheeler[37]. This further underpins the simulated universe concept and offers possible explanations to a number of unsolved problems in physics, as well as complementing and expanding our understanding of all branches of physics, the universe, and its governing laws.

In this chapter, we discuss in more detail why information is not simply a mathematical construct and can be considered fifth form of matter along with solid, liquid, gas, and plasma.

4.1 Entropy of the information bearing states

Entropy encapsulates the intrinsic measure of disorder, randomness, or uncertainty within a system. Its significance reaches far beyond the confines of physics, permeating fields such as chemistry, biology, and information theory. The exploration of entropy unveils not only the intricate interplay between energy and matter, but also the subtle limitations imposed by the second law of thermodynamics, a principle that governs the direction of natural processes.

The origin of the concept of entropy can be traced back to the 19th century, when Rudolf Clausius and Lord Kelvin independently sought to comprehend the efficiency of heat engines. Clausius, in particular, is credited with coining the term "entropy" and formulating the fundamental equation $\Delta S = Q/T$, where ΔS represents the change in entropy, Q denotes heat transferred, and T signifies temperature. This equation embodies the intuitive notion that heat naturally flows from higher to lower temperature regions, rendering systems more disordered in the process.

The second law of thermodynamics emerges as a pivotal principle governing the behaviour of macroscopic systems. It states that the total entropy of an isolated system can never decrease over time, remaining either constant or increasing. This law, often expressed in terms of heat engines, establishes a fundamental directionality for natural processes. While isolated fluctuations might lead to local reductions in entropy, the overall trend invariably points towards a state of greater disorder and higher entropy.

The implications of the second law are profound. It elucidates why heat flows spontaneously from hot to cold regions, why mixing substances tends to lead to a more disordered state, and why energy transformations inherently involve some loss as heat. In essence, the second law solidifies the arrow of time by delineating the natural progression from ordered to disordered states.

As thermodynamics matured, it became clear that entropy was a central player in elucidating the perplexing behaviour of natural systems. Boltzmann's groundbreaking work further connected entropy to statistical mechanics, revealing a profound link between the microscopic behaviour of particles and the macroscopic properties of substances. Boltzmann's entropy equation links the entropy of a system, S to the number of possible microstates, Ω compatible with the system's macrostate:

$$S = k_b \cdot \text{h } \Omega \tag{4.1}$$

where $k_b = 1.38064 \times 10^{-23}$ J/K is the Boltzmann constant. In fact, Boltzmann was so proud of his equation that he carved it on his tombstone.

It is important to clearly distinguish between the physical entropy of a system and its information entropy. The physical entropy of a given system is a measure of all its possible physical microstates compatible with the macrostate, S_{Phys}. This is a characteristic of the non-information bearing microstates within the system (see figure 4.2). Assuming the same system, and assuming that one is able to create N information states within the same physical system (for example, writing digital bits in it), the effect of creating a number of N information states is to form N additional information microstates superimposed onto the existing physical microstates (Figure 4.2). These additional microstates are information bearing states, and the additional entropy associated with them is called the entropy of information, S_{Info}. We can now define the total entropy of the system as the sum of the initial physical entropy and the newly created entropy of information, $S_{tot} = S_{Phys} + S_{Info}$. Hence, a fundamental observation that is somehow counterintuitive is that information creation increases the entropy of a given system, as illustrated in Figure 4.2. One would be inclined to assume that information creation is a process in which things get organised, ordered, and have a given purpose, so entropy would decrease. This is not the case, and information creation actually increases the entropy of a system. The image below shows a random physical system with total entropy S_{Phys}. The system is then subjected to an information creation process, with digital data imprinted on it in this example. The result is an increase in the total entropy of the system due to the newly created information bearing states, so $S_{tot} = S_{Phys} + S_{Info}$.

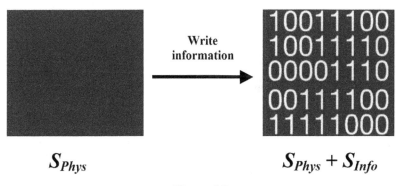

S_{Phys} $S_{Phys} + S_{Info}$

Figure 4.2

It is also important to clarify what we mean by information and information state. Throughout this book, information state is defined as any physical state, process, or event that can contain information in Shannon's information theory framework[26], and information is the amount of bit content extracted from observing such a state/event. The reader should not confuse the Shannon information entropy, IE (also known as $H(X)$), with the entropy of the information bearing states, S_{Info}. Although the two parameters are closely linked, they are rather different quantities.

If N information states are created within a given system containing U independent and distinctive information states, $N \geq U$, then the additional possible states, also known as distinct messages in Shannon's original formalism, are equivalent to the number of information bearing microstates, Ω compatible with the macrostate[35]:

$$\Omega = U^{N \cdot IE} \tag{4.2}$$

where IE is given by (3.5). Taking units of bits, $b = 2$, the general entropy of the information bearing states is now derived by combining Boltzmann's equation (4.1) with (4.2):

$$S_{Info} = N \cdot k_b \cdot \ln U \cdot \sum_{j=1}^{U} p_j \cdot \log_2 \frac{1}{p_j} \tag{4.3}$$

or $S_{Info} = N \cdot k_b \cdot \ln U \cdot IE$, where $k_b = 1.38064 \times 10^{-23}$ J/K is the Boltzmann constant.

In the case of digital data, relations (4.2) and (4.3) become

$$\Omega = 2^{N \cdot IE} \qquad (4.4)$$

$$S_{Info} = N \cdot k_b \cdot \ln 2 \cdot \sum_{j=1}^{2} p_j \cdot \log_2 \frac{1}{p_j} \qquad (4.5)$$

and in the case of genetic information, they are:

$$\Omega = 4^{N \cdot IE} \qquad (4.6)$$

$$S_{Info} = N \cdot k_b \cdot \ln 4 \cdot \sum_{j=1}^{4} p_j \cdot \log_2 \frac{1}{p_j} \qquad (4.7)$$

4.2 Landauer's principle

Landauer's principle is deeply grounded in the profound understanding that computational processes are intricately intertwined with a multitude of physical processes. By recognising this connection, we come to realise that the laws governing physical phenomena, including the principles of thermodynamics, hold sway over the domain of computational operations as well. Just as physical systems are bound by the laws that dictate the conservation of energy and the irreversibility of certain processes, so too are computational operations subject to these fundamental constraints.

Hence, a computational process that leads to the creation of digital information via some sort of physical process must also obey the laws of physics, including thermodynamics. Therefore, there must be a direct connection between the process of creating, manipulating, or erasing information and thermodynamics.

In 1961, Landauer first proposed a link between thermodynamics and information by postulating that the logical irreversibility of a computational process implies physical irreversibility[29]. Since irreversible processes are dissipative, it follows that logical irreversibility is also a dissipative process. If one considers the logical operation "erase" of a bit of digital information to be an irreversible process, then it must dissipate energy at erasure. The energy dissipation comes from the information itself, so by extrapolation, information must be physical[30].

Let's examine this argument in more detail. Digital information is stored in a memory device as a distinct, finite array of N binary elements, which can hold information without dissipation. Let us consider an isolated physical system that works as a digital memory device consisting of an array of N bits. Using (4.4) we can calculate that there are $\Omega = 2^{N \times IE}$ possible microstates, and the initial entropy of the information bearing states of the system is $S^i_{info} = N \times k_b \cdot ln(2) \times IE$.

To simplify the argument, let's assume we only have one bit of data stored, so $N = 1$.

Using (3.5) the IE of a single bit is:

$$IE = -\left(\frac{1}{2}\log_2\left(\frac{1}{2}\right) + \frac{1}{2}\log_2\left(\frac{1}{2}\right)\right) = \log_2 2 = 1$$

The meaning of $IE = 1$ is that 1 bit of information is required to encode a one letter message, or conversely, observing the above event generates 1 bit of information. For $N = 1$, $IE = 1$, then for one bit $\Omega = 2$ (i.e. 0 or 1) and the initial entropy of information of a single bit is then $S^i_{info} = k_b \cdot ln(2)$.

The erase operation can be seen as the removal of the information state, i.e. $N = 0$ or as a reset operation with all bits in the 1 or 0 state, which in the case of a single bit, means that the memory is brought to a predefined 1 or 0 state, so $IE = 0$. In both cases, the result is the entropy of the bit state being reduced to zero, either because $N = 0$ or $IE = 0$, so $S_{info} = 0$. Hence, the "erase" operation of a single bit decreases the information entropy of the system, $\Delta S_{info} = S^f_{info} - S^i_{info} = - k_b \cdot ln2$.

Since the second law of thermodynamics states that the total entropy change cannot decrease over time, $\Delta S_{tot} = \Delta S_{phys} + \Delta S_{info} \geq 0$, then the irreversible computation must reduce the information entropy of the information bearing states by increasing the entropy of the non-information bearing states via a thermal dissipation of energy:

$$\Delta Q/T = \Delta S_{phys} \geq k_b \cdot ln(2) \tag{4.8}$$

For one bit of information lost irreversibly, the entropy of the system must increase, realising an absolute value of heat per bit lost:

$$\Delta Q = k_b \cdot T \cdot ln(2) \tag{4.9}$$

Relation (4.9) is known as Landauer's principle[29][30]. It states that information is not just a mathematical construct but a physical quantity, and the minimum energy associated with a bit of information at a temperature T is given by Landauer's principle. If N bits are erased, then this energy is multiplied by a factor of N.

This implies that the energy cost of performing computations carries significant weight, and it's intriguing to consider how Landauer's principle sheds light on the minimum energy expenditure required to erase a single bit of information.

Although Landauer's principle has been the subject of some controversy, today the scientific community widely accepts it, and we refer the reader to the recent experimental confirmation of Landauer's principle[31-34], as well as various theoretical arguments in support of it[38].

It is instructive to revisit the protocol deployed to logically "erase" digital information. Performing an irreversible logical operation like "erase" brings the system into one of the three equivalent erased states shown in figures 4.3 b, c and d, for an array of 8 bits, also known as a byte. The erased state defined by Landauer is in fact a reset operation with all bits in a 1 or 0 state, but they are equivalent to a true "erased" state that is neither 0, nor 1, when the information state is fully removed from the system (Figure 4.3 c). An example of a true erased state would be an array of bits in a magnetic data storage memory, in which the erase operation does not imply the reset of all bits to an identical magnetised state but total demagnetisation of each bit, so neither 1, nor 0 could be identified in any of the bits.

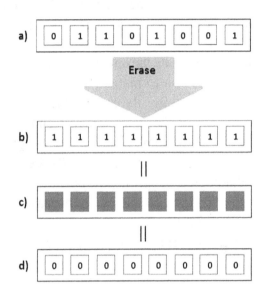

Figure 4.3

4.3 Global information power estimation

We now explore the implications of Landauer's principle in terms of global information production and in the context of the Information Catastrophe discussed in Chapter 1.

Since Landauer's principle requires a bit of information to have an intrinsic energy given by (4.9), due to the conservation of energy, an energy input of at least the same value, $k_B T \times ln(2)$, is required to create a bit of information.

In Chapter 1, we mentioned that the current annual rate of digital bit production on Earth is staggering: $N_b = 7.3 \times 10^{21}$ bits. One would naturally ask what the power constrains are to achieve such incredible volumes of digital content. Andrae and Edler from Huawei Technologies Sweden recently published an estimate of the global electricity usage that can be attributed to consumer devices, communication networks, and data centres between 2010 and 2030[39]. Their estimates showed that communication technologies could use as much as 51% of global electricity capacity by 2030.

Here we estimate the energy and power needs to sustain the annual production of information, assuming an annual growth rate of f% per year. Currently, the energy required to write a bit of information, regardless of the data storage technology used, is much higher than the minimum predicted energy, $Q_{bit} = k_B T \times ln(2) \approx 18\ meV$ at room temperature $(T = 300K)$. Let us assume that our future technological progress will allow us to write digital information with maximum efficiency. In this case, the total energy necessary to create all the digital information in a given n^{th} year, assuming f% year-on-year growth, is given by:

$$Q_{info}(n^{th}) = N_b \cdot k_B T \cdot \ln(2) \cdot (f+1)^n \qquad (4.10)$$

The total planetary power requirement to sustain digital information production is obtained by dividing relation (4.10) by the number of seconds in a year, t = 3.154 × 10⁷.

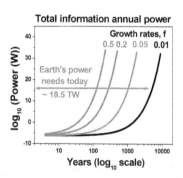

Total information annual power

Figure 4.4

Figure 4.4 shows the annual power versus number of years on a logarithmic scale, for f = 1%, 5%, 20%, and 50%, respectively. The total power requirement

48

today to power all industries, transportation, and domestic energy needs on Earth is around 18.5×10^{12} W $= 18.5$TW[40], i.e. in logarithmic value this is $\log_{10}(18.5 \times 10^{12}) = 13.27$. As could be seen from the figure, for a 1% growth rate, after ~4500 years, the creation of digital content will take up the equivalent of all of today's planetary power requirements. Similarly, for 5%, 20%, and 50% growth rates, this will occur after ~918, ~246, and ~110 years, respectively. It is worth reminding the reader that these estimates have assumed the production of digital content at maximum efficiency, which is certainly not the case yet. Hence, it appears that the current growth rate is unsustainable and that digital information production will be limited in the future by planetary power constraints. This limitation could be formulated in terms of entropy, as introduced by Deutscher in his book, The Entropy Crisis, in which he argues that the energy crisis is in fact an entropy crisis, because the entropy increase in the biosphere requires energy[41]. Since information production actually increases the entropy of a system, by extrapolation, producing digital information also increases the entropy of the biosphere. Interestingly, this increase in the information entropy of the biosphere could be used in reverse to harvest energy from entropy, as previously proposed[42].

4.4 Mass – Energy – Information equivalence principle

We established that the process of creating information requires a minimum amount of work $W \geq k_b \cdot T \cdot ln(2)$ externally applied to modify the physical system and create a bit of information, while the process of erasing a bit of information generates $\Delta Q = k_b \cdot T \cdot ln(2)$ heat energy released to the environment. However, once a bit of information is created, assuming no external perturbations, it can stay like this indefinitely without any energy dissipation. In a 2019 article, a radical idea was proposed, in which the process of holding information indefinitely without energy dissipation can be explained by the fact that once a bit of information is created, it acquires a finite mass, m_{bit} [35]. This is the equivalent mass of the excess energy created in the process of lowering the information entropy when a bit of information is erased. Using the mass-energy equivalence principle, the mass of a bit of information is derived as:

$$m_{bit} = \frac{k_b T \cdot \ln(2)}{c^2} \tag{4.11}$$

where c is the speed of light and T is the temperature at which the bit of information is stored. Having the information content stored in a physical mass allows the information to be held indefinitely without energy dissipation. Erasing the information requires external input and the mass m_{bit} is converted back into energy / heat. The implications of this rationale are that the equivalence mass – energy principle inferred from special relativity can be extrapolated to the *mass – energy – information* (M/E/I) equivalence principle, as depicted in figure 4.5.

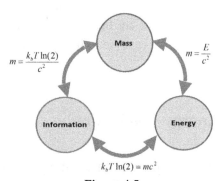

Figure 4.5

The M/E/I equivalence principle proposed in 2019 is an extension of Landauer's principle stating that, if information is equivalent to energy, according to Landauer, and if energy is equivalent to mass, according to Einstein's special relativity, then the triad of mass, energy, and information must all be equivalent, too (i.e. if $M = E$ and $E = I$, then $M = E = I$).

Furthermore, the information depends on the temperature at which the information bit exists. From (4.11), $m_{bit} = 0$ at $T = 0$ K, so as expected, no information can exist at zero absolute. Using the relation (4.11) at room temperature ($T = 300$ K), the estimated mass of a bit is $\sim 3.19 \times 10^{-38}$ Kg.

To understand this concept underpinning the M/E/I equivalence principle, let us imagine a balance as a memory device, as shown in figure 4.6 below. When the balance has no left or right tilt, i.e. it is fully balanced, the device is an erased memory state, storing no information. By convention, when it tilts to the left, the device is in memory state "1", and when it tilts to the right, it is in memory state "0". The balance will tilt only when some mechanical work is performed against it, and it will always revert to its erased state when the perturbing force is cancelled. In order to make the device hold a bit of information, a permanent force / work must be present. However, digital information requires an initial input of energy to create a bit, which is then stored indefinitely without energy dissipation. The equivalent of this process in terms of our thought balance memory device experiment is when external work is performed to place an object of finite mass on the left or right side of the balance. This is the "write" process in the memory. However, having the mass present allows a digital "1" or "0" state to be maintained indefinitely without energy dissipation. The memory erase process is equivalent to external work done to remove the mass from the balance. In this process the mass is converted back into heat, as described in the Landauer principle and confirmed experimentally.

The balance memory device imaginary experiment also shows the energy cycles corresponding to transitions from an erased state to a "1" or "0" bit of information and back from a bit of information to an erased state. A minimum energy input or output is required to transition a bit in or out of the erased state. However, once a bit of information is created, transitions from "1" to "0" and vice versa

51

can take place without the dissipation associated with this process. This is equivalent to moving the mass from the left of the balance to the right directly without going through the erased state.

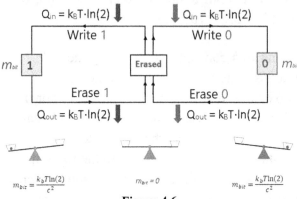

Figure 4.6

Although the M/E/I equivalence principle still awaits experimental verification, assuming it is correct, it opens up interesting possibilities with wide ranging implications for computing technologies, physics, and cosmology. Essentially, a bit of information could be seen as an abstract information particle with no charge, no spin, and a rest mass of $m_{bit} = (k_B T \times ln(2))/c^2$. In fact, it was proposed that "information" is not only the fifth form of matter along with solid, liquid, gas and plasma, but also possibly the dominant form of matter in the universe[35][36].

Although this principle was first formulated in 2019, it is important to remind the reader that these ideas are not new. The legendary physicist John Archibald Wheeler, considered the universe to be made up of three parts: particles, fields, and Information. In fact, Wheeler proposed reformulating the whole of physics in terms of information theory. He summarised his ideas in a paper that he delivered at the Santa Fe Institute in 1989[37], in which he postulated that the universe emanates from the information inherent within it and coined the phrase "It from Bit." Moreover, other scientists also independently estimated the mass of a bit of information[43][44][45][46][47].

Since the publication of this work, a small number of scientists have expressed criticism of the M/E/I equivalence principle. Their main objection revolves around the concept of information having mass, because they make a confusion between the concept of information as defined in these studies and the digital information states that underpin a bit of information in a data storage device, which are indeed physical states of matter. Digital states in data storage systems utilise some form of material support for each digital bit state. Depending on the technology deployed, this could be a bunch of magnetic nano grains forming collectively a bit, in which 0 and 1 states are defined as different magnetization states of the bit. Other technologies could utilise charged/discharged capacitors to store information, in which a charged capacitor would be a logical 1, and a discharged capacitor would be a logical 0 digital state. In any case, the bit of information has some kind of material support. Of course, in this sense, the information bit has a mass, which is the mass of the physical system that supports the bit: i.e. for instance, the mass of the nano capacitor memory cell. Another example is genetic information, which is encoded in the four letters A, C, G, and T. These represent the four nucleotides: adenine (A), cytosine (C), guanine (G), and thymine (T), and each letter can encode a maximum of two bits of information. If one takes adenine (A) for example, according to the M/E/I equivalence principle, it can have a maximum mass equivalent of two bits. However, the material support of these two bits is the adenine itself, which has the chemical formula $C_5H_5N_5$. Its mass is therefore the mass of carbon, hydrogen, and nitrogen atoms multiplied by 5. This is a much larger mass and very different from the mass of two bits of information predicted by the M/E/I equivalence principle. This is where most critics make the biggest confusion, as they do not dissociate the information bit mass from the mass of the physical support of the bit. When talking about the M/E/I equivalence principle and the mass of information, we are not referring to these material elements that support the bit. We are truly referring to the mathematical entity that defines the memory bit in probabilistic terms within Shannon's information theory. In other words, if we could master a process of creating information not supported by physical states, then these bits would have mass as described by the M/E/I equivalence principle. When material states are utilised to store information, the mass of the physical bit state is billions of times larger than the mass of the information, as derived from the M/E/I equivalence principle. Moreover, the potential energies involved in write/read out/erase processes are also multiple orders of magnitude larger than the Landauer's energies required to write/erase a memory bit.

4.5   The mass of the world's data

Here we explore the implications of the M/E/I equivalence principle, in the context of the current digital information revolution. Using the M/E/I equivalence principle, the rest mass of a digital bit of information at room temperature is m_{bit} = *3.19 × 10⁻³⁸ Kg*. We can now estimate how much information mass we are creating / converting on Earth at present, every year, as the product $N_b \times m_{bit}$. The total calculated mass of all the information we produce yearly on Earth at present is *23.3 × 10⁻¹⁷* Kg. This is extremely insignificant and impossible to notice.

For comparison, this mass is 1000 billion times smaller than the mass of a single grain of rice, or about the mass of one E.coli bacteria[48]. It will take longer than the age of the universe to produce 1 Kg of information mass if the current annual rate of data production remains the same, so *f = 0*. However, the production of digital information is rapidly increasing every year, and the objective of this work is to estimate the total information mass after a number of *n* years. Let us assume again that *f %* is the annual growth factor of digital content creation on Earth. This allows the estimation of the total information mass accumulated on the planet after *n* years of *f%* growth as:

$$M_{info}(n) = N_b \cdot \frac{k_B T \cdot \ln(2)}{f \cdot c^2} \cdot \left((f+1)^{n+1} - 1 \right)$$

(4.12)

Assuming a conservative annual growth of digital content creation of 1%, using (4.12) we estimate that it will take around ~3150 years to produce the first cumulative 1Kg of digital information mass on the planet, and it will take ~8800 years to convert half of the planet's mass into digital information mass. When

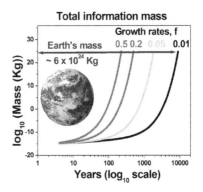

Figure 4.7

we input larger growth rates of 5%, 20%, and 50%, respectively, these numbers become extreme.

The data represented in figure 4.7 is on logarithmic scales. At a 5% growth rate of digital content production, the first 1 Kg of information mass occurs after ~675 years, and the half planetary mass is reached after ~1830 years. Similarly, for 20% and 50% growth rates, the numbers are ~ 188 years, and ~50 years, respectively for 1 Kg of information mass, and ~ 495 years, ~ 225 years for half of Earth's mass, respectively. Essentially, in the extreme case scenario where our digital information production growth is sustained at 50% per year, by the year 2070, we will have 1 Kg of digital bit content on the planet stored on all the traditional and cloud data storage centres, endpoints such as PCs, smart-phones, and Internet of Things (IoT) devices. Similarly, at 50% growth per year, by the year 2245, half of the planet's mass will be made up of digital bits.

Hence, in not a very distant future, most of the planet's mass could be made up of bits of information. Applying the law of conservation in conjunction with the M/E/I equivalence principle means that the mass of the planet is unchanged over time. However, our technological progress inverts radically the distribution of the Earth's matter from predominantly ordinary matter to the fifth form of digital information matter. In this context, assuming the planetary power limitations are solved, we could envisage a future world mostly computer simulated and dominated by digital bits and computer code. A natural question then is: *Could this already be the case?*

4.6 Is dark matter information?

The emergence of the M/E/I equivalence principle facilitates new research tools capable of addressing several scientific puzzles that continue to elude even the brightest minds of our time. One of these unsolved problems of modern physics is the nature of the mysterious substance known as "dark matter". Dark matter was first suggested in the 1920s to explain observed anomalies in stellar velocities[49], and later in the 1930s, some unseen dark matter was again required to explain the dynamics and stability of clusters of galaxies[50][51][52]. However, the strongest necessity for dark matter's presence came in the 1970s from researching the galaxy rotation curves, which are diagrams representing the orbital velocity of gas and stars in galaxies as functions of their distance to the galactic centre[53][54]. The orbital velocity of a rotating disc of gas and stars is expected to obey Kepler's second law, so that the rotation velocities will decline with distance from the centre. Experimental observations indicate that the rotation curves of galaxies remain flat as their distance from the centre increases[55][56]. Since there is more gravitational pull than expected only from the observed light/baryonic matter of a galaxy, the flat rotation velocity curves are a strong argument that dark matter should exist (see figure 4.8).

Although the existence of dark matter is generally accepted, a significant community of scientists is working on alternative explanations that do not require the existence of dark matter.

There are various theoretical approaches, but they usually involve modifications of existing established theories such as modified Newtonian dynamics, modified general relativity, entropic gravity, tensor-vector-scalar gravity, and so on[57][58][59][60][61].

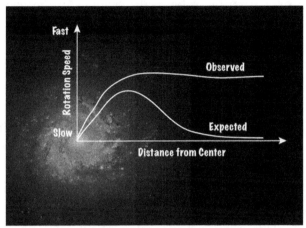

Figure 4.8

Today many physicists are trying to identify the nature of dark matter by a variety of means, but the consensus is that dark matter is composed primarily of a not yet discovered subatomic particle[62]. Unfortunately, all efforts to isolate or detect dark matter have failed so far.

Following up on the discovery of the M/E/I equivalence principle, here we propose a radical new idea that could explain the nature of dark matter in the universe. Within the concept of a simulated digital universe, then information should be present in huge amounts all over the universe, perhaps within the space-time itself. When this concept is combined with the M/E/I equivalence principle, requiring the information to have tiny physical mass, one would conclude that information should be the dominant form of matter in the universe. According to the M/E/I equivalence principle, an informational digital universe would contain a large amount of "hidden" energy/mass in the form of information. This has a baryonic mass manifestation via the gravitational interactions, but it is impossible to detect because a bit of information would have no charge, no spin, and no other properties except mass, so it would not interact with the electromagnetic radiation. These are in fact the characteristics of the elusive "dark matter", whose presence is inferred only from the rotational dynamics of the galaxies[63][64], but has never been observed or detected.

Assuming constant temperature and without making any considerations of where this information mass is localised in space-time, a rough estimation could be performed. Taking the average temperature of the universe, $T = 2.73$ K, and using (4.11), the estimated mass of a bit of information is $m_{bit} = 2.91 \times 10^{-40}$ Kg.

It is well accepted that the matter distribution in the universe is ~ 5% ordinary baryonic matter, ~27% dark matter, and ~68% dark energy[22]. This implies that there is about 5.4 times more dark matter than visible ordinary matter. Taking the estimated mass of our Milky Way galaxy as ~ 7×10^{11} $M_\odot$ solar masses[65], and using the mass of the sun, $M_\odot \sim 2 \times 10^{30}$ Kg, then the estimated dark matter mass in our galaxy is $M_{Dark\_Matter} \sim 3.78 \times 10^{12} M_\odot = 7.56 \times 10^{42}$ Kg. Assuming that all the missing dark matter is made up of bits of information, then the entire Milky Way galaxy has N_{bits} (Milky Way) = $M_{Dark\_Matter} / m_{bit}(T=2.73\text{K}) = 2.59 \times 10^{82}$ bits. The estimated number of galaxies in the visible universe is ~ 2×10^{12} [66], so the estimated total number of bits of information in the visible universe is ~ 52×10^{93} bits.

In Chapter 1, we estimated the total number of particles in the universe to be around ~ 10^{80}, which is a factor of 10^{13} smaller than the number of bits required to make up the missing dark matter. Hence, it is reasonable to assume that information could be stored in elementary particles, similar to DNA, but perhaps this would not be enough to make up the missing matter, and the search for the missing information mass should continue.

4.7 The DNA of matter

The issue of information content associated with physical measurement, matter and, by extension, the information content of the universe has been a topic of scientific debate since the late 1920s. Leo Szilard analysed the relationship of information to physical processes in 1929[67], demonstrating that information about a system dictates its possible ways of evolution and behaviour, and offering an elegant solution to Maxwell's famous paradox, known as the Maxwell's Demon[68].

With the emergence of digital computers, digital technologies, and digital data storage, the topic of information physics entered a new era. In 1961, Rolf Landauer first proposed a link between thermodynamics and information, demonstrating that digital information is in fact physical[29]. Later, Landauer speculated that the universe could be a giant computer simulating itself[69], an idea further developed in more recent studies by Lloyd[19][70][71]. Wheeler considered the universe to be made up of particles, fields, and information, proposing that everything emanates from the information inherent within it, i.e. "It from Bit"[37]. Other studies extrapolated the physical nature of information to estimate the mass of information[43-47], culminating with the recent publication of the mass-energy-information equivalence principle[35], and Vopson's postulate that "information" is the fifth form of dominant matter in the universe along with solid, liquid, gas, and plasma[5][36].

These radical theories are based on the principle that information is physical, the information is registered by physical systems, and all physical systems can register information[72]. Accordingly, there is a given amount of information stored in the universe, regardless of whether it is observed or not. The proposed existence of this information imposes some fundamental questions about it:

"Why is there information stored in the universe, and where is it?"
and
"How much information is stored in the universe?"

Let us deal with these questions in detail. To answer the first question, let us imagine an observer tracking and analysing a random elementary particle. Let us assume that this particle is a free electron moving in the vacuum of space, but the observer has no prior knowledge of the particle or its properties. Upon tracking the particle and commencing the studies, the observer will determine, via meticulous measurements, that the particle has a mass of *9.109 × 10^{-31} kg*, charge of *– 1.602 × 10^{-19} C* and a spin of *1/2*. If the examined particle was already known or theoretically predicted, then the observer would be able to match its properties to an electron, in this case, and confirm that what was observed / detected was indeed an electron. The key aspect here is the fact that by undertaking the observations and performing the measurements, the observer did not create any information. The three degrees of freedom that describe the electron, any electron anywhere in the universe, or any elementary particle were already embedded somewhere, most likely in the particle itself. This is equivalent to saying that particles and elementary particles store information about themselves, or, by extrapolation, that there is information content stored in the matter of the universe, similar to "matter DNA". Due to the mass-energy-information equivalence principle[35], we postulate that information can only be stored in particles that are stable and have a non-zero rest mass, while interaction / force carrier bosons can only transfer information via waveform. Hence, in this work, we are only examining the information content stored in the matter particles that make up the observable universe, but it is important to mention that information could also be stored in other forms, including on the surface of the space-time fabric itself, according to the holographic principle[73].

The second question is: *"How much information content is there in the observable universe?"*

This question has been addressed in several studies going back as far as late 1970s. Using the Bekenstein – Hawking formula for black-hole entropy[74][75], Davies calculated the information content of the universe by applying the black-hole entropy formula to the whole universe[76]:

$$I \approx \frac{2\pi G M_u^{\,2}}{hc} = 10^{120}\, bits \tag{4.13}$$

where G is the gravitational constant, M_u is the mass of the universe enclosed within its horizon, h is Planck's constant, and c is the speed of light. In fact, Davies produced a time dependent information formula that allows studying the time evolution of the information in the universe, as discussed in detail by Treumann[77].

Wheeler estimated the number of bits in the present universe at $T = 2.735\ K$ from entropy considerations, resulting in 8×10^{88} bits content[37]. Lloyd took a similar approach and estimated the total information capacity of the universe as[70]:

$$I = \frac{S}{k_B \ln(2)} \approx \left(\rho c^5 t^4 / \hbar\right)^{3/4} \approx 10^{90}\, bits \tag{4.14}$$

where S is the total entropy of the matter dominated universe, k_B is the Boltzmann constant, ρ is the matter density of the universe, t is the age of the universe at the present time, c is the speed of light, and $\hbar$ is the reduced Planck constant.

Applying Landauer's principle, Gough developed an information equation of state and calculated that 10^{87} bits of intrinsic universe information content could account for all the dark energy[78]. His estimates are for a cosmic time coinciding with star formation, when most of the universe's bit information content is represented by high temperature baryons, giving an average bit energy value of 120 eV [78]. Using the mass-energy-information equivalence principle[35], Vopson estimated that around 52×10^{93} bits would be enough to account for all the missing dark matter in the observable universe[35][36]. However, this is most likely overestimated, as he

61

assumed that all the information bits are stored at $T = 2.735\ K$. This is inaccurate because a significant amount of baryonic matter is contained in stars, intergalactic gas, and dust, which all have temperatures larger than the CMB.

The prime objective of this study is to estimate the bits of information contained in the matter particles of the observable universe via a new approach using Shannon's information theory. This approach is therefore independent of the dynamics of the expansion of the universe, or the internal dynamics of the universe, giving an accurate bit content at any given time of observation.

4.8 The information content per elementary particle

We will use Shannon's information entropy to estimate the information content of matter in the universe. The observable universe is made up of building blocks called elementary particles. According to the Standard Model, there are 30 elementary particles and antiparticles known today. These are six quarks and their corresponding anti-particles, six leptons and their corresponding anti-particles, five vector bosons, and one scalar boson. Out of all these particles, only a fraction of them are participants in the observable universe, as observed today. Most of the other particles or anti-particles are unstable and have extremely short lifetimes, so their observation is only possible under artificially created experimental conditions or theoretically. Therefore, their contribution to the observable universe is negligible, and, by extrapolation, their capacity to register information is also negligible. As already postulated in the previous section, all force carrier particles/bosons and particles with zero rest mass cannot store/register information and they can only transfer information, so they are not considered in our estimates.

Each elementary particle is fully described by at least three degrees of freedom: mass, charge, and spin. As we already discussed, these degrees of freedom are embedded within each elementary particle, like an intrinsic label, or "particle DNA". We regard this as the information content registered in each particle about itself, measured in bits. Excluding the bosons, the anti-particles, and all the unstable elementary particles, out of all 30 elementary particles, only protons, neutrons, and electrons are actual participants in the observable universe today. Of course, protons and neutrons are in fact not elementary particles, as they are made up of three quarks each, which would need to be accounted for.

In Section 1.2 we already estimated the number of particles in the universe. Taking into account the internal structure of the atoms by using the exact number of p^+, e^- and n^0 in each kind of atom and the % abundance of all atoms in the universe, we estimated the probabilities of observing a p^+, e^- and n^0 in the observable universe as: $P_{p+} = P_{e-} = 0.466$ and $P_{n0} = 0.067$. We can now regard this baryonic distribution of matter in the observable universe as a set of three possible states / events, $\{p^+, e^-, n^0\}$, with the occurrence probabilities $\{P_{p+}, P_{e-}, P_{n0}\} = \{0.466, 0.466, 0.067\}$. Shannon's classical information theory gives us an elegant mechanism to estimate the bits of information content per event. In the case of a three-state system, the maximum information (I) encoded per event is $IE = log_2 3 = 1.585$ bits, corresponding to the case when the three events / states are equally probable. This is not the case here, so the average bit information content per event, meaning per any element of the set $\{p^+, e^-, n^0\}$, is given by:

$$IE_{pen} = -\sum_{j=1}^{n} P_j \cdot \log_2 P_j = -\left(P_{p+} \log_2 P_{p+} + P_{e-} \log_2 P_{e-} + P_{n0} \log_2 P_{n0}\right) \qquad (4.15)$$

Introducing the numerical values in (4.15), we obtain $IE_{pen} = 1.288$ bits of information encoded per elementary particle. The meaning of this result is that the most effective way that information can be compressed into a p^+, e^- or n^0, or the average amount of information content stored in each of these particles, is 1.288 bits per particle.

To find out the total information stored in all the matter particles in the universe, one needs to multiply the total number of protons, electrons, and neutrons in the observable universe, i.e. 4×10^{80}, by the number of bits of information per elementary particle, i.e. 1.288.

Let us remember that the information content estimated above is in fact inaccurate, as protons and neutrons are not elementary particles and have an internal structure containing three quarks each. According to the standard model, a proton contains two up quarks and one down quark, and a neutron contains one up quark and two down quarks. Hence, we can estimate the number of up / down quarks per each kind of atom, and using the % abundance of atoms, we can estimate how many up and down quarks are in a sample of 100 atoms. Summing up, we obtain a total of 216.55 up quarks and 108.27 down quarks contained in protons, and 15.6 up quarks and 31.21 down quarks contained in neutrons. This results in a total of 232.15 up quarks and 139.48 down quarks. Adding up the total number of electrons (108.27) to the total number of up quarks and down quarks, we obtain 479.9, which is the total number of elementary particles contained in a statistical sample of 100 random atoms. Using the acronyms *uq* for up quarks and *dq* for down quarks, we can now regard, again, these elementary particles as a set of three possible states / events, where the three possible states are $\{uq, dq, e^-\}$, and their occurrence probabilities are $\{P_{uq}, P_{dq}, P_{e^-}\}$. The occurrence probabilities are obtained as $P_{uq} = 232.15 / 479.9$, $P_{dq} = 139.48 / 479.9$ and $P_{e^-} = 108.27 / 479.9$, resulting in $\{P_{uq}, P_{dq}, P_{e^-}\} = \{0.483, 0.29, 0.225\}$.

Shannon's information entropy formula for this set of events and probabilities is:

$$IE_{uqdqe} = -\sum_{j=1}^{n} P_j \cdot \log_2 P_j = -\left(P_{uq} \log_2 P_{uq} + P_{dq} \log_2 P_{dq} + P_{e^-} \log_2 P_{e^-}\right) \quad (4.16)$$

Introducing the numerical values in (4.16) we obtain $IE_{uqdqe} = 1.509$ bits of information encoded per elementary particle. If the number of particles in the universe, N_b, is known, as is the information content per elementary particle in the observable universe, the total information stored in the baryonic matter in the universe could be worked out as $N_{bits} = N_{tot} \times IE_{uqdqe} = 6.036 \times 10^{80}$ bits.

This value is lower than the previous estimates. However, our calculation refers only to the information stored in particles such as electrons, protons, neutrons, and their constituent quarks. The previous estimates refer to the total information content of the universe. Moreover, our calculation also takes a different approach, using information theory, which gives the most effective information compression. The discrepancy in the information content values could therefore be explained by the fact that our estimate is for the most effective compression mechanism, according to information theory. Other possible reasons for the information content discrepancy could perhaps suggest that the universe most likely contains more information, which is stored in other elementary particles, or other media, not accounted for in this study. For example, we have considered all bosons to be force / interaction particles responsible for the transfer of information, rather than the storage of information, which might not be the case. We also ignored all the anti-particles, as well as all neutrinos. Although there is scope for further refinements of this work, this is the first time that information theory has been used to estimate the information content of the observable universe, and this formalism would most certainly would stimulate future studies. Moreover, the current approach offers a unique tool for estimating the information content per elementary particle, which is very useful for designing practical experiments to test these predictions.

5

Proposed Experiments for Testing the Information Conjectures

The information conjectures stating that: a) information is physical and has mass; b) information is the fifth form of matter in the universe; and c) the concept of "matter DNA", are truly transformational and open up incredible research avenues in all branches of science. However, as long as these remain in the theoretical realm, their validity will always be questioned and debated. The ultimate proof can only come from the experimental confirmation of these conjectures. To be considered valid, an experimental confirmation must be also reproducible by other research laboratories, be easily repeatable, and be able to rule out any other possible explanations or experimental artefacts. Once these conditions are fulfilled, assuming the experiment is positively confirmed, then the scientific community should accept the theory as valid. In this chapter, we are exploring possible experimental protocols that could be deployed to prove or disprove the information conjectures.

5.1 Mass of data

In what follows, we propose a simple experiment capable of testing this theory by physically measuring the mass of digital information. This consists of an ultra accurate mass measurement of a digital data storage device when all its memory bits are in a fully erased state. This is then followed by the operation of recording digital data on all of its memory bits until it is at full capacity, followed by another accurate mass measurement. If the proposed mass – energy – information equivalence principle is correct, then the data storage device should be heavier when information is stored on it than when it is in a fully erased state. One could easily estimate the mass difference, Δm in this experiment. Let us assume a memory device with 1 Tb of storage capacity, then the total number of memory bits is 10^{12} bytes = 8×10^{12} bits, as 1 byte = 8 bits.

Using the relation (4.11) at room temperature ($T = 300K$), the estimated mass of a bit is ~ 3.19×10^{-38} Kg. Hence, the predicted mass change in this experiment is $\Delta m = 2.5\times10^{-25}$ Kg. The proposed experiment is simple in terms of physical complexity, but very challenging overall, as success depends on one's ability to accurately measure mass changes in the order of ~10^{-25} Kg. The required measurement sensitivity could be reduced by a factor f if the amount of data storage under test is increased from 1Tb to $f \times 1Tb$. One measurement option would be a sensitive interferometer similar to the Laser Interferometer Gravitational-Wave Observatory (LIGO), although smaller sizes and sensitivities would be sufficient for the proposed measurement. More feasible to test the proposed principle is to use an ultra-sensitive Kibble balance used for defining the Kilogram, such as the one developed at the National Physical Laboratory in the UK.

However, this proposal remains unachievable today as the current measurement sensitivities are off by many orders of magnitude from what is required to perform the experiment.

5.2 Mass of hot objects

The recent prediction of the information mass content per elementary particle, allows us to extend the range of possible experiments beyond digital data storage, to a simple material body of mass m. Because the mass of information is temperature

68

dependent, in this experiment one could simply confirm the information conjectures by observing the effect of the temperature change on the information mass content of elementary particles contained within a physical body of a known mass. Let us consider a random mono-atomic solid of mass m, made up of identical atoms of atomic mass weight A, each atom containing N_{e-} electrons, N_{p+} protons, and N_{n0} neutrons. If each elementary particle contains I bits of information, then a mass m would contain N_b bits of information:

$$N_b = I \cdot \frac{mN_A}{A}\left(N_{e^-} + 3\left(N_{p^+} + N_{n_0}\right)\right) \tag{5.1}$$

where N_A is Avogadro's number, $N_A = 6.022 \times 10^{23}$ mole^{-1}, and the factor 3 accounts for the fact that each proton and each neutron is made up of three quarks. According to the M/E/I equivalence principle, for a temperature change ΔT, the general expression of the information mass change Δm^{inf} of a body of mass m is:

$$\Delta m^{inf} = I \cdot \frac{mN_A k_b \Delta T \ln(2)}{Ac^2}\left(N_{e^-} + 3\left(N_{p^+} + N_{n_0}\right)\right) \tag{5.2}$$

where $k_b = 1.38064 \times 10^{-23}$ J/K is the Boltzmann constant and c is the speed of light. Relation (5.2) predicts a temperature dependence of the information mass change. Hence, one could design an experiment to measure the information mass change inflicted by a temperature change on a body of mass m. Since the physical mass of the material under test does not change with temperature (assuming solid materials are thermally and chemically stable), the detected mass change can only be related to the information mass change, providing a direct confirmation of the proposed information conjectures.

Let us assume a metallic body of $m = 1$ Kg copper (Cu), with each Cu atom containing $N_{e-} = 29$ electrons, $N_{p+} = 29$ protons, and $N_{n0} = 34.5$ neutrons. The fractional value of N_{n0} accounts for the existence of the two Cu isotopes containing 34 neutrons (70%) and

36 neutrons (30%), respectively. This proportion of isotopes gives a relative atomic mass number of $A = 63.55$ g. If each subatomic elementary particle contains $I = 1.509$ bits of information as predicted previously, then using (5.1) we obtain the total number of bits of information stored in a Kg of Cu as $N_b = 29.8 \times 10^{26}$ bits. For a temperature change $\Delta T = 100$ K of the Cu sample (cooling or heating), using (5.2) we obtain an absolute value of information mass change of $\Delta m^{inf} = 3.33 \times 10^{-11}$ Kg. Although this value significantly improves the required measurement resolution relative to the initial proposed experiment ($\Delta m^{inf} \sim 10^{-25}$ Kg), accurate measurement of $\sim 10^{-11}$ Kg is still extremely challenging.

5.3 Detection of information at erasure

A third experimental protocol combines the estimate of the information content per elementary particle, with the M/E/I equivalence principle, to formulate a new way of testing the information conjectures.

In order to explain the rationale behind this proposed experiment, let us first consider an elementary particle. For convenience, we will consider an electron. We also assume that the electron stores I_{e^-} bits of information about itself. According to the M/E/I equivalence principle, the electron's rest mass is the sum of its physical mass and its information mass:

$$m_{e^-} = m_{e^-}^{phys} + m_{e^-}^{inf} \tag{5.3}$$

Although here we are examining an electron, this conjecture applies to any elementary particle that is stable and has a non-zero rest mass. The mass of a bit at temperature T is given by:

$$m_{bit} = \frac{k_b T \ln(2)}{c^2} \tag{5.4}$$

Hence, the mass of the electron becomes:

$$m_{e^-} = m_{e^-}{}^{phys} + \frac{I_{e^-} k_b T \ln(2)}{c^2} \tag{5.5}$$

A quick numerical estimate indicates that the information mass of the electron is extremely small, so that:

$$\frac{m_{e^-}}{m_{e^-}{}^{inf}} = \frac{9.11 \times 10^{-31}}{I_{e^-} \cdot 3.19 \times 10^{-38}} \approx \frac{2.85}{I_{e^-}} \times 10^7 \tag{5.6}$$

Taking $I_{e^-} = 1.288$ bits, it results that the rest mass of the electron is ~ 22 million times larger than its information mass, indicating that indeed, the mass of the electron is well approximated by its physical rest mass, while its information mass is negligible. Again, this makes experimental testing impossible via direct mass change measurements.

Instead, we propose an experiment that involves the measurement of the information mass indirectly, via an information erasure process. According to the M/E/I equivalence principle and Landauer's principle, the information mass must be dissipated as energy upon erasure.

How can one erase the information contained within an electron?

In order to completely erase the information within any elementary particle, one needs to remove the particle from existence. This can be achieved via a matter – antimatter annihilation reaction. Luckily, in the case of an electron, there is a routinely accessible process known as electron – positron annihilation, where the positron (e⁺) is the antiparticle of the electron (e⁻), and a collision between an electron and a positron can lead to their mutual annihilation. In the annihilation process, the rest mass energies and the kinetic energies of the electron and positron are converted into radiation. Depending on the total spin of the positron - electron pair, the annihilation process can take place via two possible pathways for the emitted radiation. When the total spin is one, the annihilation

71

produces three gamma photons. When the positron - electron pair has a total spin of zero, the annihilation process produces two gamma photons.

This latter process is an ideal candidate to study the information content of the input particles by examining what may arise from the erasure of the information upon their annihilation. The total energy of the colliding electron – positron pair is:

$$E_{tot} = E_{e^-} + K_{e^-} + E_{e^+} + K_{e^+} \tag{5.7}$$

where $E_{e^-} = m_{e^-}c^2$ and $E_{e^+} = m_{e^+}c^2$ are the rest mass energies of the electron and positron, respectively. $K_{e^-} = m_{e^-}v_{e^-}^2/2$ and $K_{e^+} = m_{e^+}v_{e^+}^2/2$ are the kinetic energies of the colliding particles moving with velocities v_{e^-} and v_{e^+}. Since $c \gg v_{e^-}$ and v_{e^+}, the kinetic energies are negligible. Performing the experiment with a beam of slow positrons and static electrons can easily meet this condition. The electron – positron pair must conserve the total energy, the momentum, and the angular momentum after the annihilation process. Energy conservation ensures that two 511 keV gamma photons are produced from the conversion of their rest mass energies. The conservation of angular momentum is automatically fulfilled in the two-photon annihilation process, where one spin is up and the other spin is down. The momentum conservation imposes the condition that these two gamma photons travel in opposite directions to each other. Reaction (5.8) and figure 5.1 a) show the standard electron – positron annihilation process that produces two gamma photons.

$$m_{e^-}c^2 + m_{e^+}c^2 = \gamma + \gamma \tag{5.8}$$

72

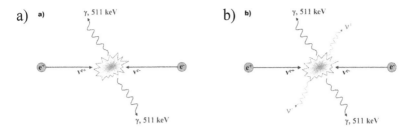

Figure 5.1

However, relation (5.7) does not include any information energy that might be contained in the particles themselves. Accounting for the information content and neglecting the kinetic energies, the total energy is:

$$E_{tot} = m_{e^-}c^2 + m_{e^+}c^2 + I_{e^-}k_bT\ln(2) + I_{e^+}k_bT\ln(2) \qquad (5.9)$$

where $T_{e^-} = T_{e^+} = T$ because the positrons will reach thermal equilibrium with the metallic sheet containing the target electrons, so each particle will have the same temperature at the time of collision. I_{e^-} and I_{e^+} are the amount of information stored by the electron and the positron, respectively. The rest masses of the electron and positron, as well as their information contents, must be equal to each other. Energy conservation ensures again that two gamma photons of about 511 keV are produced. However, if particles store information, upon annihilation (i.e. erasure), the information content must also be conserved by producing two information energy photons, v^+ and v^-.

The conservation of momentum imposes the condition that these two additional photons also travel in opposite directions to each other. Reaction (5.10) and figure 5.1 b) show the electron – positron annihilation process that includes the information erasure.

$$m_{e^-}c^2 + m_{e^+}c^2 + I_{e^-}k_bT\ln(2) + I_{e^+}k_bT\ln(2) = \gamma + \gamma + v^+ + v^- \;(5.10)$$

5.3.1 Theoretical predictions

The successful detection of the information energy photons ν^+ and ν^- will confirm both information conjectures: i) the mass-energy-information equivalence principle; ii) the bit information content of elementary particles, implying the existence of information as the 5th state of matter.

The information energy photons have very specific characteristics that allow their identification with a high degree of confidence. Firstly, they should emerge simultaneously with the 511 keV gamma photons. This means that synchronised detection of the gamma and the information energy photons would offer a strong indication of their origin.

Secondly, the information energy photons have very specific wavelengths, which are proportional to the amount of information stored by the electron and the positron, but also proportional to their temperature.

In the previous section, we estimated the information content per elementary particle to be 1.509 bits. Although the estimation accounted only for stable elementary particles excluding anti-particles, we can assume that the information content of the positron is equal to that of the electron, so $I_{e^-} = I_{e^+} = I = 1.509$ bits, and upon erasure, the resulting photons also have the same energies/frequencies $\nu^+ = \nu^- = \nu$.

According to the M/E/I equivalence principle, the wavelength of the information energy photons is:

$$\lambda = \frac{hc}{Ik_b T \ln(2)} \tag{5.11}$$

where $h = 6.62 \times 10^{-34} \, m^2 \, kg/s$ is the Planck's constant. The predicted wavelength of the information energy photons as a function of the temperature, which extends from mid-infrared (MIR) to far-infrared (FIR) spectral regions, is shown in figure 5.2. Since the information content of 1.509 bits is a theoretical prediction that has not been confirmed yet, it is instructive to extend the possible range of the bit information content per particle.

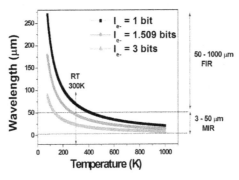

Figure 5.2

Hence, we also computed the predicted values for two additional information contents of 1 and 3 bits per particle, respectively (see figure 5.2). The data shows that, for 1.509 bits of information content, the expected information energy photon wavelength ranges from 3 to 180 μm, depending on the temperature of the experiment. At room temperature, information energy photons of ~50 μm wavelength are predicted to emerge. This value changes proportionally to the bit information content, so from 1 to 3 bits per elementary particle, the wavelength at room temperature ranges broadly from 25 to 75 μm. Knowing these predicted values is very important for the experimental design and the selection of suitable IR detectors.

5.3.2   Proposed experimental design

The experiment should be designed to ensure that not only the two 511 keV gamma photons are detected, but also the additional two IR photons, ν^+ and ν^-. The detection of the IR photons presents some additional challenges because they are easily attenuated within the sample. For our experiment, we propose to use positrons generated by a ^{22}Na radioactive source.

Figure 5.3 shows the decay scheme of ^{22}Na. This isotope is very convenient because of its relative low cost, long half-life of 2.6 years, and high positron yield. Positrons emitted via nuclear radioactive decay of ^{22}Na sources have an energy distribution range from 0 to 545 keV, and 90.4% of the time they decay according to the following reaction:

$$^{22}_{11}Na \rightarrow ^{22}_{10}Ne + e^+ + v + \gamma \tag{5.12}$$

where e+ is the positron, n is a neutrino, and γ is a 1274 keV gamma photon. Unfortunately, the γ-decay reaction of ^{22}Na generates high-energy positrons (also known as fast positrons) and they have a large penetration range into the sample material.

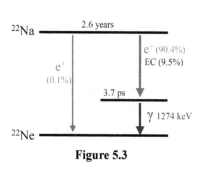

Figure 5.3

Therefore, the sample material must be thick enough to absorb the positrons, but thin enough to ensure it does not attenuate the 511 keV gamma rays that are created in the electron – positron annihilation within the sample. Most importantly, we need to ensure that the two IR photons v^+ and v^- produced at the erasure of the information content, are also not fully attenuated within the sample material. In order to fulfil these requirements, one option is to use a metallic thin layer target material bombarded with low energy positrons (also known as slow positrons). Slow positrons have a higher probability of electron annihilation, as they diffuse through the target material. When fast/high-energy positrons enter a material, they lose energy by interacting with it, slowing down to thermal energies. This thermalization process takes only a few picoseconds, while the positron mean lifetime in metals ranges from 100 to 450 picoseconds[79]. Fast-to-slow positron moderation is easy to accomplish using a moderation step made of a suitable material that has a negative work-function for positrons[80][81].

Fast positrons penetrate the moderation step, where some will emerge on the other side as fast positrons but with reduced energies, some will annihilate inside the moderator, and some will thermalize and diffuse to reappear at the surface of the moderation step where they are spontaneously emitted as mono-energetic slow positrons of kinetic energy close to the work function of the moderation material (a few eV). The fast-to-slow positron conversion efficiency is typically $\sim 10^{-4}$ [82] and one of the best moderation materials is tungsten single-crystal[83][84][85][86]. We propose to cover the ^{22}Na source with a thin (1 - 2 μm) single-crystal tungsten foil in (100) orientation[84], which has a negative work-function of around 3 eV.

Alternatively, a polycrystalline thin film tungsten moderator[85] could be coated directly onto the ^{22}Na positron source via a suitable thin film deposition process. The low energy positrons leaving the moderator will annihilate inside the metallic target material. The range of the positrons in the target dictates the choice of metal. For example, 545 keV positrons penetrating an Al target have a mean lifetime of 166 ps and a range of 0.954 mm, while for an Au target the mean lifetime is 118 ps and the range is 0.194 mm[87], which is almost 5 times shorter than that of Al. To ensure a high probability of positron – electron annihilation, we propose to use a metallic Al thin film as the target material. The thickness of the Al thin film must be in the range of a few nm so the thermalized positrons can undergo surface annihilation in the Al material and a large fraction of the resulting photons (gamma and IR) can reach the detectors.

Figure 5.4 shows a schematic diagram of the proposed experiment (thickness of the layers is not at scale)[88].

It is important that the W moderator and the Al thin film sample, completely encapsulate the ^{22}Na source. The experimental design requires one end of the ^{22}Na source to be in contact with a temperature controller, while the other surface is used for the positron beam. Gamma and IR detectors are placed in close proximity to the Al thin film. The temperature controller (cooling and heating) is required because the theory predicts that the IR photons detected at information erasure display a linear temperature scaling of their energy.

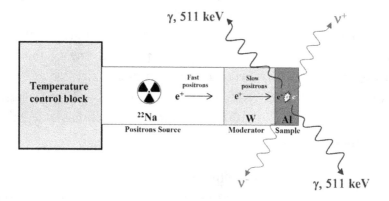

Figure 5.4

Assuming the successful detection of the IR information energy photons, the ability to vary the temperature of the sample will act as a double confirmation of the experiment by detecting the wavelength change of the IR photons with the temperature. This experimental geometry is the easiest path to controlling the temperature of the emitter (positrons) and sample (electrons), simultaneously. However, if the infrared photons are totally absorbed in the Al sample, then a different, more complex experimental geometry could be designed, in which the Al film and the W moderator are detached from the source.

In order to detect the IR information photons with the maximum degree of confidence, we need to rule out the detection of background IR photons, that are present everywhere, especially at the frequencies expected for the IR information photons. This could be done effectively by taking advantage of the fact that the IR information photons are emitted simultaneously with the 511 keV gamma rays. Since gamma photons are produced at collisions with positrons, one way of achieving this detection effectively is to modulate the positron beam using an electro-optical beam chopper. This in turn allows the deployment of lock-in detection. The lock-in detection technique is a powerful method used in various scientific and engineering applications to extract weak signals from noisy backgrounds. This technique enhances the signal-to-noise ratio by modulating the input signal with a chopper and then demodulating it using synchronous detection. By modulating the input signal at a known frequency and employing phase-sensitive detection, the lock-in technique exploits the correlation between the modulating reference signal and the input signal to significantly amplify the signal of interest while minimising the impact of noise. The electro-mechanical chopper serves as the modulator in this technique. It consists of a rotating blade that intermittently blocks and allows the transmission of the input signal, which in our case would be the beam of positrons. Hence, the chopper generates a modulated signal where the amplitude alternates between high and low levels at a specific modulation frequency.

The reference signal has the same frequency as the chopper modulation and is then fed into a lock-in amplifier for the synchronous detection process. The second signal fed into the lock-in amplifier is the IR detector signal. A low-pass filter is employed to filter out the high-frequency components, including noise, leaving only the components at the modulation frequency. Synchronous detection is the heart of the lock-in detection technique. The filtered signal from the low-pass filter represents the amplified signal of interest. By tracking the phase of the reference signal, the lock-in amplifier can extract the in-phase (I) and quadrature (Q) components of the filtered signal.

79

The proposed lock-in detection technique with signal modulation of the positron beam offers huge advantages, such as exceptional sensitivity and selectivity, helping to discern extremely weak signals from overwhelming noise backgrounds. By deploying this technique, the IR photons detected would be categorically the information photons we are searching for.

This experiment was first proposed in 2022[88]. So far, no research group has managed to construct the instrument or perform the measurements. In fact, the author of this book attempted to source the funding for this experiment via a crowd-funding campaign in late 2022[89]. Despite the amazing support received from over 100 backers, the crowd-funding campaign has only achieved around 2.5% of the funding required to construct the instrument. This book is another attempt to secure funding for this experiment, as all the sale proceeds will go to fund the proposed experiment.

6

Second Law
of Information Dynamics

One of the most powerful laws in physics is the second law of thermodynamics, which describes the evolution of entropy over time. In fact, the second law is applicable to the evolution of the entire universe, and Clausius stated, "The entropy of the universe tends to a maximum".

Examining the time evolution of information systems, defined as physical systems containing information states within Shannon's information theory framework, in 2022, a new fundamental law of physics has been proposed and demonstrated, called the second law of information dynamics, or simply the second law of infodynamics[90]. Its name is an analogy to the second law of thermodynamics, which describes the time evolution of the physical entropy of an isolated system, which requires the entropy to remain constant or to increase over time. In contrast to the second law of thermodynamics, the second law of infodynamics states that the information entropy of systems containing information states must remain constant or decrease over time, reaching a certain minimum value at equilibrium. Mathematically, the second law of infodynamics states that:

$$\frac{\partial S_{Info}}{\partial t} \leq 0 \qquad\qquad (6.1)$$

where S_{info} is the general entropy of the information bearing states, given by equation (4.3) and rewritten here for clarity:

$$S_{Info} = N \cdot k_b \cdot \ln U \cdot \sum_{j=1}^{U} p_j \cdot \log_b \frac{1}{p_j} \qquad (6.2)$$

where N is the total number of information states within a given system containing U independent and distinctive information states, $N \geq U$.

Examining (6.1) and (6.2), the decrease in the entropy of the information states can only come from the reduction over time in the total number of states, N, or a reduction over time in the Shannon entropy, due to changes to the probabilities p_j. This is because all the other terms are constants. These important aspects will be examined in more detail in this chapter, including empirical exemplification using real world systems. However, uncovering this surprising new law of physics has massive implications for all branches of science and technology. With the ever-increasing importance of information systems such as digital information storage or biological information stored in DNA / RNA genetic sequences, this new physics law offers an additional tool for examining these systems and their time evolution.

Additionally, perhaps the most important implication of the second law of infodynamics appears to be the fact that its behaviour points to the characteristics of a computational system, underpinning to some degree the simulated universe hypothesis.

In this chapter, we will examine the second law of infodynamics in greater detail and demonstrate the universal nature of this new physics law.

6.1 The necessity of the second law of infodynamics

Our universe can only be in two possible states: finite/close or infinite/open. Although there is no definite evidence for either case, the current scientific consensus is that we live in an infinite universe that is in continuous expansion. What is important, though, is the fact that, regardless of whether the universe is finite or infinite, the thermodynamic laws are equally applicable anywhere in the universe, to any process within it. The first law of thermodynamics states that energy can neither be created, nor destroyed. The energy is conserved. The energy in the universe can only be converted from one form to another, but overall, it remains constant. Using Clausius' sign convention, the mathematical differential form of the first law of thermodynamics is:

$$dQ = dU + dW \qquad (6.3)$$

where Q is the net heat energy supplied to the universe, W captures the work done by the universe in all possible forms, and U represents the total internal energy of matter and radiation in the universe.

However, the universe does not exchange heat with anything. If the universe is infinite/open then there is nothing to exchange heat with because there is nothing outside the universe. If the universe is closed/finite, it could be seen as being contained within something larger, so theoretically it could exchange heat with something. However, if the universe is finite/closed as in a simulated construct, then the universe does not exchange heat with anything. We therefore start from the assumption that the universe does not exchange heat with anything, regardless of its state: finite or infinite. For a universe that is expanding adiabatically, the first law becomes:

$$0 = dQ = dU + dW \qquad (6.4)$$

We now recall the relation that links heat to entropy, $dQ = T.dS$, where S is the total entropy of the universe and T is the temperature. Since T has a non-zero value as dictated by the third law of thermodynamics, and the average temperature of the observable universe could in fact be considered to be 2.7 K, we deduce that *dS = 0*. This implies that the total entropy of the universe must be constant. This constant entropy does not violate the second law of thermodynamics, which allows the entropy to be constant over time, or to increase. However, in an expanding universe, the entropy will always increase because more possible microstates are being created via the expansion of the space itself / universe. Figure 6.1 shows a diagram of a physical system containing matter, when the size of the system is in continuous expansion, while its physical content remains unchanged. Just as in our expanding universe, the space expansion in the schematic physical system shown above facilitates the emergence of more microstates, and the total entropy increases rapidly. If the universe does not expand, at some point the entropy will reach its maximum and the universe will achieve equilibrium.

This process allows for the physical entropy of the universe in the past to have been at its maximum value, when the universe would have been much smaller than today and at near equilibrium.

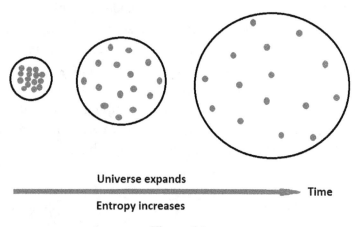

Universe expands

—————————————————————————————► Time

Entropy increases

Figure 6.1

The evidence for this is the cosmic microwave background (CMB) radiation[91], which is almost isotropic, having a temperature of ~2.7 K in all directions[92], and a very low-level temperature anisotropy ($\Delta T / T \sim 10^{-5}$)[93][94]. The time origin of this low level of temperature anisotropy can be traced back to ~370,000 years after the big bang[95], when the universe was close to chemical and thermal equilibrium and the density inhomogeneities were comparable to the temperature anisotropies ($\Delta\rho/\rho \sim \Delta T / T \sim 10^{-5}$).

However, in order to comply with the first law of thermodynamics and the adiabatic expansion, we just showed that the total entropy of the universe must be constant. If this is the case, how can the physical entropy of our expanding universe increase continuously? This is called the "Entropic Paradox" and to solve it, there are only three possibilities:

a) The laws of thermodynamics are not valid;

b) The universe is not expanding;

c) The entropy budget of the universe contains an unaccounted entropy term;

The readers would agree that possibilities a) and b) are out of the question as these are supported by undisputed empirical evidence. We are therefore left with the search for another entropy term responsible for the initial high entropy of the universe. This entropy term must also balance the total entropy budget of the universe in order to ensure that the overall entropy remains constant over time, despite the evident increase in physical entropy that we can observe in the expanding universe.

We propose that the missing entropy term is the entropy associated with the information content of the universe, or the entropy of the information states within the universe.

Let's write the total entropy of the universe, S as the sum of the physical entropy and the information entropy:

$$S = S_{Phys} + S_{Info} \tag{6.5}$$

Differentiating (6.5) we get:

$$dS = dS_{Phys} + dS_{Info} \qquad (6.6)$$

Imposing the dS = 0 condition, and taking a time derivative, we obtain:

$$\frac{dS_{Phys}}{dt} + \frac{dS_{Info}}{dt} = 0 \qquad (6.7)$$

Since $dS_{Phys}/dt \geq 0$, i.e. physical entropy always increases over time according to the second law of thermodynamics and according to empirical observations, then the increase in the physical entropy must be balanced by the decrease in the information entropy over the same time interval, so $-\frac{dS_{Phys}}{dt} = \frac{dS_{Info}}{dt}$, which means:

$$\frac{dS_{Info}}{dt} \leq 0 \qquad (6.8)$$

Relation (6.8) is identical to relation (6.1), and it is exactly the mathematical expression of the second law of infodynamics, requiring that the entropy of the information states must remain constant or decrease over time. Hence, the second law of infodynamics appears to be a cosmological necessity. It is important to realise that in order for the overall entropy of the universe to remain constant, the absolute values of physical entropy and information entropy do not have to be equal. Only their rate of change over time must be equal, in order to ensure a constant overall entropy of the universe.

We would like to return briefly to the "Entropic Paradox" introduced here. This must not be confused with a different "Entropic Paradox" that is related to the fact that the entropy of the universe since the big bang is still increasing, indicating that the universe started from a very low entropy state. However, cosmological observations such as the cosmic microwave background radiation indicate that the universe must have been at equilibrium at some point in the past. That would mean its entropy was at its maximum. Since the entropy was at its maximum, how is it possible to have a continuous entropy increase in the universe? This is also coined the "Entropic Paradox". However, in this section, we discussed this issue and explained how a system that is at equilibrium and maximum entropy can evolve towards another maximum entropy (see figure 6.1). This is due to the expansion of the system (i.e the universe expansion), which creates more possible microstates for the same amount of matter and energy in it. This is a process by which a system at its maximum entropy and equilibrium state could evolve in such a way that its entropy still increases, and therefore it is a false paradox.

6.2 Second law of infodynamics and digital information

Let us assume a digital data storage system containing a total of N information states and two distinct states, $U = 2$ (i.e. 0 and 1). This can be any type of digital data storage technology, but in this study we will focus on a magnetic data storage system. According to (4.2) the system will have a total number of possible microstates:

$$\Omega = 2^{N \cdot IE} \tag{6.9}$$

From (4.3), the entropy of the information bearing states for this digital information system is:

$$S_{Info} = N \cdot k_b \cdot \ln 2 \cdot \sum_{j=1}^{2} p_j \cdot \log_2 \frac{1}{p_j} \tag{6.10}$$

The maximum Shannon information entropy of this system, $IE = \sum_{j=1}^{2} p_j \cdot \log_2 \frac{1}{p_j}$, is $IE = 1$. Although the IE can deviate slightly from this upper limit, for large N it is reasonable to assume that the IE is stable over time and $IE \rightarrow 1$.

Hence, in the case of digital information, the only parameter that can drive the time evolution of the entropy of the information bearing states is the total number of states, N. If N increases, then the information entropy increases. However, there is no mechanism that would result in spontaneous information being created without external intervention (i.e. energy input). The only possible evolution of N over time is down or constant, in accordance with the second law of infodynamics.

In the case discussed here, the N information states are physically determined by magnetization changes in the material. For any given non-zero Kelvin temperature T, these magnetic states will undergo magnetic relaxation processes with a relaxation time (τ) given by the well-known Arrhenius-Neel equation:

$$\frac{1}{\tau} \cong v_0 e^{-\frac{K_a V}{k_b T}} \tag{6.11}$$

where v_0 is the magnetization attempt frequency to overcome the energy barrier, which is approximately $v_0 \approx 10^9$ Hz, K_a is the anisotropy constant of the magnetic material, V is the magnetic grain volume and k_b is the Boltzmann constant. The meaning of this relaxation time is the average time it takes for a magnetic grain of volume V within a magnetic bit state to undergo a spontaneous magnetization flip due to thermal activation. Hence, after a sufficiently long time, we expect magnetic grains to lose their magnetization state, leading to magnetic bit states undergoing self-erasure, and therefore reducing the information states N. The implication of this is that the entropy of information bearing states tends to decrease over time. This is a very straightforward process and also a direct consequence of the second law of thermodynamics because, over time, the digital states are eroded by thermal fluctuations, leading to the self-erasure of data. The higher the temperature of the environment, the more probable the data self-erasure processes are. Hence, in the case of digital information, the second law of infodynamics is rather trivial and fully expected.

To demonstrate this, we simulated a granular magnetic thin film structure with perpendicular uniaxial anisotropy of $K_a = 8.75 \times 10^7$ J/m^3, saturation magnetization $M_s = 1710$ kA/m and average unit cell size (cubic) $V = 10^{-27}$ m^3, at room temperature $T = 300$ K. A standard cell size volume suitable for magnetic recording should be sufficiently large to maintain the thermally stable magnetization of the cell for ~10 years (i.e. τ in equation (6) is 3.15×10^8 seconds). Under this condition, the ratio of magnetiocrystalline energy to thermal energy at $T = 300$ K should be around 40, $\frac{K_a V}{k_b T} \approx 40$, resulting in a cell size volume of $V \approx 1.9 \times 10^{-27}$ m^3. The unit cell size volume in our simulations has been intentionally taken 1.9 times lower in order to speed up the computation time. These values resulted in a relaxation time of 1.5 s, which corresponds to a single iteration in the Monte Carlo algorithm[96]. The simulated thin film sample size was 400nm × 550nm × 2nm, giving a bit size of 50nm × 50nm. Starting with a thermalized random state, INFORMATION is suddenly written using the digital binary code. Using a micromagnetic Monte Carlo algorithm[96], we tracked the information loss as the system is allowed to thermalize over a period of time. The data indicates that the system evolves over time in such a way that the second law of thermodynamics is indeed fulfilled by the physical entropy and the total entropy of the system. However, when the entropy of the information bearing states is examined independently, we conclude that the second law manifests in reverse, so that the information entropy stays constant or decreases.

Figure 6.2 a) shows schematically the word INFORMATION written on a material in binary code using magnetic recording. Red denotes magnetization pointing out of the plane, and blue is magnetization pointing into the plane; Figure 6.2 b) – d) shows the time evolution of the digital magnetic recording information states simulated using Micromagnetic Monte Carlo. b) Initial random state; c) INFORMATION is written (t = 0s); d) Iteration 930 (t = 1395s) showing the degradation of information states.

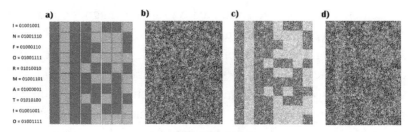

Figure 6.2

The simulation performed on our test sample resulted in a simultaneous reduction of the magnetization of all the magnetic information bit states up to the point when $N = 0$. However, in reality, this process can take place gradually so that N reduces to a lower value in steps until it eventually reaches zero.

The time evolution of the information entropy of digital information confirms the second law of infodynamics.

6.3 Second law of infodynamics and genetic information

The DNA or RNA genetic molecule is a highly complex information encoding system that displays incredible properties such as an extremely low error rate, high stability, the ability to self-replicate and self-repair, and the ability to produce other molecules. The biological information is encoded in the sequence of nucleotides and the time evolution of a genome is described by the occasionally induced genetic mutations. Mutations are errors that appear during the process of self-replication when the genome is damaged and left unrepaired. Genetic mutations manifest as changes in the genetic sequence and can take place via three mechanisms: i) Single nucleotide polymorphisms (SNPs), where changes occur so that the number of nucleotides N remains constant; ii) Deletions, where N decreases; iii) Insertions, when N is increasing. Genetic mutations can affect an organism's phenotype and are essential for the diversity and evolution of living organisms.

Because of protective mechanisms like DNA / RNA self-repair or proofreading during replication, mutation rates are very low, but avoiding mutations could become too energetically costly for living organisms, and they never reach zero. The scientific consensus on the dynamics of genetic mutations is the Darwinian theory of evolution, in which genetic mutations are random processes[97]. This means that mutations occur randomly, regardless of whether an organism will benefit or not from the DNA/RNA changes. Only natural selection determines which mutations are beneficial and preserved in the course of evolution[98], but there is no observed correlation between any parameter or variable and the probability that these mutations will occur or not.

Here we demonstrate that genetic mutations are actually driven by a hidden information entropic force, governed by the second law of infodynamics. This observation is very powerful because it challenges the Darwinian view that genetic mutations are complete random events, and it could be used to develop predictive algorithms of genetic mutations before they occur[99].

To demonstrate this we recall that DNA/RNA sequences encode biological information in a long string of the letters A, C, G, and T, where the characters represent adenine (A), cytosine (C), guanine (G), and thymine (T) (replaced with uracil (U) in RNA sequences). Therefore, within Shannon's information theory framework, a typical genome sequence can be represented as a probabilistic system of $U = 4$ distinctive states, $X = \{A,C,G,T\}$ and probabilities $p = \{p_A, p_C, p_G, p_T\}$. Using digital information units and equation (3.5), we determine that the maximum Shannon information entropy of a given genome is $IE = 2$, so each nucleotide can encode a maximum 2 bits: $A = 00$, $C = 01$, $G = 10$, $T = 11$. According to (4.2), for a given genomic sequence containing N nucleotides, the total number of possible microstates is:

$$\Omega = 4^{N \cdot IE} \qquad (6.12)$$

91

According to (4.3), the entropy of the information bearing states of a genomic sequence is:

$$S_{Info} = N \cdot k_b \cdot \ln 4 \cdot \sum_{j=1}^{4} p_j \cdot \log_2 \frac{1}{p_j} \tag{6.13}$$

The time evolution of the entropy of genetic DNA/RNA information systems is given by the time evolution of their genetic mutations. Similar to the case of digital information, a reduction of N would most likely result in a reduction of the overall entropy of the information bearing states, so "deletion" mutations would automatically fulfil the second law of infodynamics.

Hence, an interesting test case would be a genomic system that undergoes frequent SNP mutations in a short period of time, maintaining the N constant. Hence, in such a system, the reduction of the entropy of the information bearing states would come not from a reduction in N, as was the case for digital information, but from a reduction in Shannon's *IE* term. An excellent genomic sample that fits these requirements is a virus genome, and, for this study, we selected the RNA sequence of the novel SARS-CoV-2 virus, which emerged in Dec. 2019 and resulted in the COVID-19 pandemic. This choice was dictated by the huge scientific interest in the COVID-19 pandemic, which led to an explosion of studies and genetic data on SARS-CoV-2 being publicly available in a short time frame.

The reference RNA sequence of the SARS-CoV-2 was collected in Wuhan, China in December 2019 (MN908947) [100] and it has 29,903 nucleotides, so $N = 29,903$. For this reference sequence we computed the Shannon information entropy using relation (3.5). The value obtained represents the reference Shannon information entropy at time zero, before any mutations took place. Using the *National Centre for Biotechnology Information (*NCBI) database, we searched and extracted a number of SARS-CoV-2 variants sequenced at various locations around the globe, at different times, starting from Jan. 2020 to Oct. 2021 (Table 6.3.1).

Table 6.3.1 Summarised results of the analysis performed on selected SARS-CoV-2 variants sequenced at various locations around the globe, over a period of 22 months.

Genome	Ref.	SNPs	Time (months)	Location	Shannon IE
MN908947	[100]	0	0	China	1.9570243
LC542809	[101]	4	3	Japan	1.9569197
MT956915	[102]	7	5	Spain	1.9569230
MW466798	[103]	9	7	South Korea	1.9569327
MW294011	[104]	19	10	Ecuador	1.9567058
MW679505	[105]	25	14	USA	1.9566630
MW735975	[106]	26	14	USA	1.9565714
OK546282.1	[107]	32	16	USA	1.9565675
OK104651.1	[108]	40	20	Egypt	1.9564591
OL351371.1	[109]	49	22	Egypt	1.9562614

By searching for complete genome sequences, containing the same number of nucleotides as the reference sequence, we carefully selected variants that displayed an incremental number of SNP mutations with time, and we computed the Shannon information entropy for each variant. The calculations have been performed using previously developed software, GENIES[110][111], designed to study genetic mutations using Shannon's information theory[28].

The full data set, including genome data references/links, collection times, the number of SNP mutations, and the Shannon information entropy value of each genome, is shown in Table 6.3.1. Figure 6.3 shows the time evolution of the number of SARS-CoV-2 SNP mutations and the time evolution of each variant's information entropy, S_{info} computed using (6.13) and normalised to k_b. The data indicates that, as expected, the number of mutations increases linearly as a function of time. Remarkably, for the same dataset, the Shannon information entropy (*IE*) and the overall information entropy of the SARS-CoV-2 variants (S_{info}) computed using (6.13), decrease rather linearly over time. The observed correlation between the information entropy and the time dynamics of the genetic mutations is truly unique, because it reconfirms the second law of infodynamics, but it also points to a possible deterministic approach to genetic mutations, currently believed to be just random events. The existence of an entopic force that governs genetic mutations instead of randomness is very powerful, and it could lead to the future development of predictive algorithms for genetic mutations before they occur.

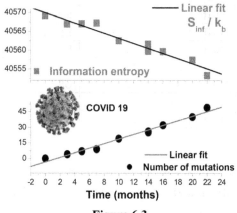

Figure 6.3

We should acknowledge that, while all analysed SARS-Cov-2 variants showed a decrease in their information entropy as they underwent genetic mutations, the data points presented in the above figure have been carefully selected to emphasise the linear trend.

Naturally, we ask next whether the same RNA system would display behaviour consistent with the second law of infodynamics, when the SARS-CoV-2 variants suffered "addition" mutations, so the number of nucleotides N is no longer constant, but becomes larger than 29,903, increasing the information entropy. Using the NCBI database, we searched all the sequenced SARS-CoV-2 variants from 1st of Jan. 2020 to 1st of Jan 2022. We searched only complete sequences with no missing/undetermined nucleotides, and the result was a total of 4.48 million sequences. When we restricted the results to only the sequences that had at least 29,903 nucleotides or more, then 48,450 sequences were identified. Unfortunately, only one suffered a mutation where the resultant number of nucleotides increased by 1 to 29,904. Hence, 98.92% of all mutations took place via "deletion" reducing the total number of nucleotides. Since only one genome out of 4,48 million appeared to increase the number of nucleotides, this is statistically irrelevant. Hence, we conclude that, for this test case, genetic mutations appear to take place in a way that reduces their information entropy, mostly via a deletion mechanism or a SNP. This is fully consistent with the second law of infodynamics, as a deletion would automatically decrease the total information entropy, and the SNPs have been shown to take place in such a way that the information entropy is again reduced due to a reduction in Shannon's information entropy, *IE*.

To support our results, we would also like to quote the famous Spiegelman's experiment that took place in 1972[112]. In this experiment, Spiegelman studied the evolution of a virus over 74 generations. The virus was kept isolated in ideal conditions to survive, and with each generation, the virus was sequenced. The initial virus had 4500 base points, and with each generation the genome decreased consistently in size. After 74 generations, the virus evolved to only 218 base points, showing an interesting and unexplained reduction of its genome of over 95%. Just as the present study on SARS-CoV-2 [90][99], Spiegelman's experiment is fully consistent with the second law of infodynamics, which requires the information entropy to remain constant or to decrease over time, reaching a minimum value at equilibrium.

6.4 Second law of infodynamics and Hund's Rule

Electronic states in atoms are fully described by four principal quantum numbers: a) the principal quantum number, *n*. This number determines the energy of a particular shell or orbit, and it takes non-zero positive integral values $n = 1, 2, 3, 4...$; b) the orbital angular momentum quantum number, ℓ. This quantum number describes the subshell, and gives the total angular momentum of an electron due to its orbital motion. This quantum number takes integral values restricted to $\ell = 0, 1, 2, ..., n - 1$; c) The magnetic quantum number, m_ℓ. This quantum number determines the component (projection) of the orbital angular momentum along a specific direction, usually the direction of an applied magnetic field. It takes integral values, and for a given value of ℓ, it may have $(2\ell + 1)$ possible values: $m_\ell = \ell, \ell\text{-}1, \ell\text{-}2, ...0, -\ell, ...-(\ell\text{-}1), -\ell$; d) The spin quantum number s, and the secondary spin quantum number, m_s. The spin quantum number s gives the eigenvalues of the spin angular momentum operator, and it is related to the fact that the electron has an intrinsic angular momentum called "spin" or spin angular momentum, which results from the rotation of the electron around an internal axis. The spin quantum number takes the values $s = n/2$, where n is a positive integer, so that $s = 0, 1/2, 1, 3/2, 2, ...$ The secondary quantum spin number m_s determines the direction (i.e. projection) of the spin angular momentum along the direction of an applied magnetic field. The allowed values of m_s are $2s + 1$ values from $-s$ to $+s$ in steps of 1. For example, an electron has $s = 1/2$, so the allowed values of m_s are $-1/2$ and $+1/2$.

The electrons occupy atomic shells according to Pauli's exclusion principle[23], which states that two or more identical fermions cannot simultaneously occupy the same quantum state within a quantum system. In the case of electrons in atoms, this means that it is impossible for two electrons in a multi-electron atom to have the same values of the four quantum numbers described above. For example, if two electrons reside in the same orbital, then their n, ℓ, and m_ℓ values are the same, so their m_s must be different, imposing that the electrons must have opposite half-integer spin projections of 1/2 and $-1/2$. In a previous section of this book we already discussed the fact that Pauli's exclusion principle, in its insistence on the distinguishability of fermions, bears an uncanny resemblance to the rules of coding and programming, being a possible manifestation of the simulation's underlying code.

In this section, we are examining another rule in atomic / subatomic physics, called Hund's rule, within the same context. Hund's rule is necessary because, despite Pauli's exclusion principle, in the case of multi-electron atoms, multiple electron arrangements are still possible while fulfilling Pauli's exclusion principle. This breaks the distinguishability condition of the electrons and additional rules appear to be responsible for this problem.

In order to determine the electron population of an atomic orbital corresponding to the ground state of a multi-electron atom, German physicist Friedrich Hund formulated in 1927 a set of rules[113] derived from phenomenological observations. These are called Hund's rules and when used in conjunction with Pauli's exclusion principle, they are useful in atomic physics to determine the electron population of atoms corresponding to the ground state.

To explain this, let's assume that an atom has 3 electrons on its p orbital. Figure 6.4 shows the three allowed ground state distinctive configurations that fulfil Pauli's exclusion principle, resulting in total spin quantum values of 1/2, 3/2 and 1/2, respectively.

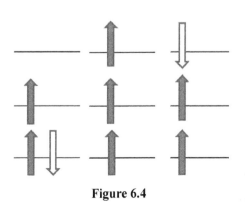

Figure 6.4

The correct electronic arrangement is given by Hund's first rule, which is the most important, and it is simply called Hund's Rule. This states that the lowest energy atomic state is the one that maximises the total spin quantum number, meaning simply that the orbitals of the subshell are each occupied singly with electrons of parallel spin before double occupation occurs. Therefore, the term with the lowest energy is also the term with the maximum number of unpaired electrons, so for our example shown in figure 6.4, the Hund's Rule dictates that the correct configuration is the middle one, resulting in a total spin quantum value of 3/2.

Hund's Rule is derived from empirical observations, and there is no clear understating of why the electrons populate atomic orbitals in this way. So far, two different physical explanations have been given[114]. Both explanations revolve around the energetic balance of the electrons and their interactions in the atom. The first mechanism implies that electrons in different orbitals are further apart, so that electron–electron repulsion energy is reduced. The second mechanism claims that the electrons in singly occupied orbitals are less effectively screened from the nucleus, resulting in a contraction of the orbitals, which increases the electron–nucleus attraction energy[115].

Here we take a novel approach and examine the electronic population in atoms within the framework of information theory. We are able to demonstrate that Hund's Rule (Hund's first rule) is a direct consequence of the second law of information dynamics. This requires that, at equilibrium in the ground state, electrons occupy the orbitals in such a way that their information entropy is minimum, or equivalently, the bit information content per electron is minimum.

6.4.1 Numerical calculations

We treat the two possible values of the secondary quantum spin number m_s of the electrons in atoms, $m_s = -1/2, +1/2$, as two possible events, or as a two letter message within Shannon's information theory framework. The secondary quantum spin number m_s is a very important parameter because it is the only quantity that distinguishes two electrons residing in the same orbital, since their n, ℓ, and m_ℓ values are the same, their m_s must be different to fulfil Pauli's exclusion principle.

We will allocate to the two possible projections of the m_s the spin up $\downarrow$ and spin down $\downarrow$ states. In this context, the set of $n = 2$ independent and distinctive information states are $X = \{\uparrow, \downarrow\}$, with a discrete probability distribution $P = \{p\uparrow, p\downarrow\}$.

Hence, for any N electrons, we have $N\uparrow$ and $N\downarrow$ electrons, so that $N = N\uparrow + N\downarrow$, and relation (3.5) gives the Shannon information entropy per electron spin, or the bit information content stored per electron spin, while relation (4.3) gives the total entropy of information states per N electrons. Hence, relation (3.5) becomes:

$$IE = p_\uparrow \cdot \log_2 \frac{1}{p_\uparrow} + p_\downarrow \cdot \log_2 \frac{1}{p_\downarrow} \tag{6.14}$$

where $p\uparrow = N\uparrow / N$ and $p\downarrow = N\uparrow / N$, which allows re-writing (6.14) as:

$$IE = \frac{N_\uparrow}{N} \cdot \log_2 \frac{N}{N_\uparrow} + \frac{N_\downarrow}{N} \cdot \log_2 \frac{N}{N_\downarrow} \tag{6.15}$$

Since the electronic populations are stable, then N is constant, and the minimum in the entropy of the information bearing states, S_{Info} corresponds to a minimum in Shannon's information entropy. We now consider the s, p, d and f orbitals and we analyse in detail the Shannon's information entropy of each possible distinctive electronic configuration, for any possible occupancy number of these orbitals. The maximum allowed value for the information entropy is 1 bit, $IE = 1$ and the minimum possible value is 0 bits. We will demonstrate that for each orbital, the configuration that has the lowest Shannon information entropy, i.e. the lowest bit information content, corresponds to the highest total spin quantum value. Hence, Hund's Rule is in fact a consequence of the second law of infodynamics.

6.4.2   s – orbital

The s orbital can accommodate a maximum of $N = 2$ electrons. Figure 6.5 shows the possible electronic configurations of an s orbital. When $N = 1$, or $N = 2$, the $IE = 1$ bit in both cases, while the total spin quantum value is 0.5 and 0, respectively. Since there are no other possible configurations, the case for s-orbital

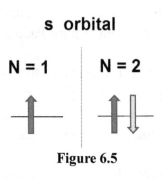

Figure 6.5

is rather trivial. Figure 6.7 a) shows a plot of the IE values versus the total spin quantum value, S for the s orbital.

6.4.3   p – orbital

The p orbital can accommodate a maximum of $N = 6$ electrons. Figure 6.6 shows the electronic populations on the p orbital for all possible N values. We should mention that only distinct configurations have been represented in the diagram. Any electronic arrangement that results in the same ratio of spin-up and spin-down electrons is not represented, as it would duplicate the results.

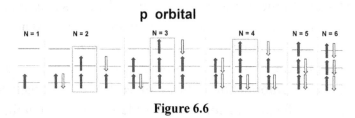

Figure 6.6

Similarly, configurations obtained by inverting all spins, i.e. mirror images, result in the same IE values and are not considered to avoid duplications.

Figure 6.7 b) shows the graph of the IE values versus the total spin quantum value for all possible distinct occupancy cases of the p orbital. As can be seen in figure 6.6, each time multiple arrangements are possible, as is the case *for N = 2, 3* and *4*, respectively, the maximum spin quantum value corresponds to the minimum *IE* value estimated using (6.15). For *N = 2, 3* the minimum *IE* is 0 in each case, while for *N = 4* the minimum *IE* value is 0.811. To emphasise this, we highlighted in figure 6.6 the correct configurations that are required by Hund's Rule.

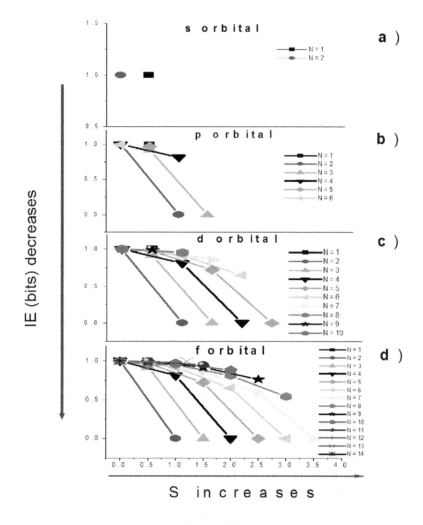

Figure 6.7

6.4.4   d – orbital

We now examine the case of the d orbital, which can accommodate a maximum of $N = 10$ electrons. Figure 6.8 shows the distinct electronic populations allowed on the d orbital for all possible N values.

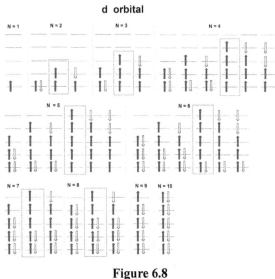

Figure 6.8

For $N =1$, 9 and 10, only a single distinct arrangement is possible, while for all the other N values, multiple electronic arrangements are allowed within Pauli's exclusion principle.

Figure 6.7 c) shows the *IE* values versus the total spin quantum value for all possible distinct occupancy cases of the d orbital. The data indicates that the maximum spin quantum value corresponds to the minimum *IE* value estimated using (6). For each N value, we highlighted in figure 6.8 the correct configurations that are required by Hund's Rule. These all correspond exactly to the lowest *IE* value, reinforcing the validity of the second law of infodynamics. The minimum *IE* value of 0 is achieved for $N = 2$, 3, 4 and 5. The minimum *IE* values are $IE = 0.65$ for $N = 6$, $IE = 0.863$ for $N = 7$ and $IE = 0.954$ for $N = 8$, respectively.

6.4.5   f – orbital

Finally, we examine the f-orbital, which can accommodate a maximum of $N = 14$ electrons. Therefore, we have 14 possible groups, with $N = 1, 13$ and 14 having only a single distinct electronic arrangement possible, while for all the other N values, multiple electronic arrangements are allowed by Pauli's exclusion principle. Figure 6.9 shows the distinct electronic populations allowed on the f orbital for all possible N values. Again, we highlighted the correct arrangements as dictated by Hund's Rule, and we calculated the IE values for all possible configurations.

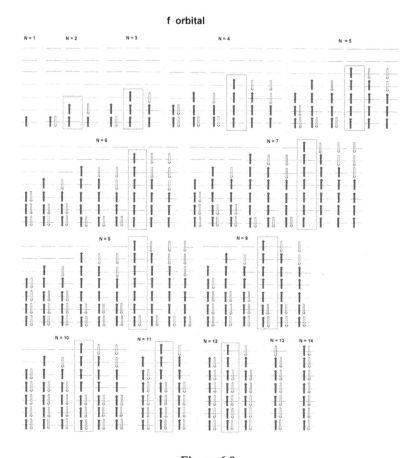

Figure 6.9

The minimum *IE* value of 0 is achieved for $N = 2, 3, 4, 5, 6$ and 7. For the remaining groups with multiple electronic configurations, the minimum *IE* values are $IE = 0.544$ for $N = 8$, $IE = 0.764$ for $N = 9$, $IE = 0.881$ for $N = 10$, $IE = 0.946$ for $N = 11$, and $IE = 0.98$ for $N = 12$, respectively. Figure 6.7 d) shows the *IE* values versus the total spin quantum value for all possible distinct occupancy cases of the f orbital. Again, the data shows categorically that in all cases, the minimum *IE* value corresponds to the maximum spin quantum value, *S*, so the second law of infodynamics appears to be the real driving force behind Hund's Rule.

6.5 Second law of infodynamics and symmetries

Symmetry is a mathematical concept in which a certain property, for instance, the geometrical shape of an object, is preserved under certain transformations applied to the object. Such transformations include translations, rotations, reflections, and more complex operations combining these. In each case, the object remains invariant upon transformation.

In the context of Euclidian geometry, these transformations are called symmetry operations. A symmetry operation is the movement of an object into an equivalent and indistinguishable orientation that is carried around a symmetry element.

A symmetry element is a point, line, or plane about which a symmetry operation is carried out.

Mathematically, let's assume an n-dimensional (nD) geometry, where an nD figure is defined as any subset of the R^n space. An isometry of the R^n space is a function $f : R^n \rightarrow R^n$ that preserves distances, so for all $x_1, x_2, ...x_n \rightarrow R^n$, the distance between any $f(x_i)$ and $f(x_j)$ is equal to the distance between x_i and x_j. Therefore, mathematically, the symmetry of an nD figure F is an isometry mapping F into itself, so $f : R^n \rightarrow R^n$, such that $f(F) = F$. For a plane image, $n = 2$, so we have a 2D geometry. Figure 6.10.shows schematically the 2D isometry.

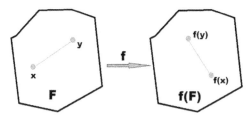

Figure 6.10

The classical group theory is the mathematical tool for the study of symmetry, describing the structure of transformations that map objects to themselves exactly.

However, symmetry is not merely a mathematical concept. It transcends disciplines, connecting mathematics, chemistry, biology, and physics, and appears to be a fundamental property of the universe.

This is evidenced by everything around us, from the elegant symmetrical patterns of snowflakes to the fundamental symmetries governing subatomic particles. Symmetry occurs at all scales, playing a pivotal role in the structure and behaviour of matter in the universe. Figure 6.11 shows a few examples of amazing symmetries manifesting in nature.

In fact, many biochemical processes are governed by symmetry, resulting in a wealth of biological structures that exhibit strong symmetries or regularity patterns.

One of the most profound connections between symmetry and physics arises from Noether's theorem, which states that for every continuous symmetry in a physical system, there is a corresponding conserved quantity. In other words, the laws of conservation, including the law of energy conservation, have their roots in the symmetries of physical systems. Symmetry principles are also at the heart of the Standard Model of particle physics, which describes the fundamental particles and their interactions.

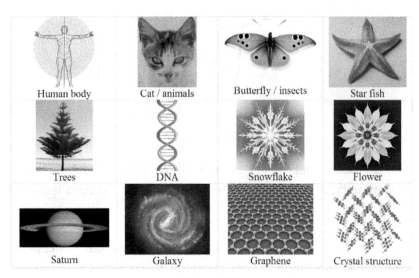

Figure 6.11

This abundance of symmetry in the natural world begs the question: *Why does symmetry dominate all systems in the universe instead of asymmetry?* After all, the entropic evolution of the universe tends to a high entropy state, yet everything in nature appears to prefer high symmetry and a high degree of order.

Here we explore the mathematical underpinnings of symmetry and its crucial significance in the realms of information theory and physics. We will demonstrate a unique observation that a high symmetry corresponds to a low information entropy state. This is exactly what the second law of infodynamics requires: systems in the universe tend to evolve over time in a way that minimises their information entropy at equilibrium. Hence, this remarkable observation appears to explain why symmetry dominates in the universe: it is due to the second law of information dynamics.

Before we proceed to our proof, it is useful to establish a way of measuring the symmetry of an object quantitatively. In other words, how much symmetry does a shape have?

106

One accepted method of measuring the symmetry of an object is by counting the number and type of symmetry operations that one can carry out on the object. The more symmetry operations a shape has, the more symmetric it is.

Since the symmetry operations are carried out around the symmetry elements, we propose to quantify the symmetry of a shape by counting its number of symmetry elements instead of counting the number of symmetry operations.

For example, a perfect square has eight symmetry operations (four rotations around the central axis and four reflections) and five symmetry elements (one axis of rotation and four axes of reflection).

Our main objective is to describe the relationship between the symmetry of an object, determined by the number of its symmetry elements (SE), and its information entropy (IE).

In order to do this, let's consider a range of simple Euclidian 2D geometric shapes. We start with an ordinary triangle, defined by three sides of length a, b, and c and three corresponding angles α, β, and γ respectively (see figure 6.12). These parameters are a unique representation of the shape, as there is no other possible way of forming from this set a triangle that looks different from the original one.

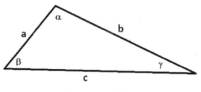

Figure 6.12

Within Shannon's information theory framework, we define the set of six characters, $N=6$, with six distinct characters, $U=N=6$, $X = \{a, b, c, \alpha, \beta, \gamma\}$, and a probability distribution $P = \{p_a, p_b, p_c, p_\alpha, p_\beta, p_\gamma\}$ on X. The probabilities of the set are:

107

$P = \left\{\frac{1}{6}, \frac{1}{6}, \frac{1}{6}, \frac{1}{6}, \frac{1}{6}, \frac{1}{6}\right\}$. The average information per character, or the number of bits of information per character for this set, is given by (3.5), $IE = -\sum_{j=1}^{6} p_j \cdot \log_2 p_j = \log_2 6 = 2.585$

This ordinary triangle has no symmetry, and accordingly, it has zero symmetry elements, so $SE = 0$.

We now examine an isosceles triangle, as shown in figure 6.13. This shape has one symmetry element (a reflection axis), so $SE = 1$ and it is fully defined by the set of four distinct characters $U = 4$, $X = \{a, b, \alpha, \beta\}$, and a probability distribution $P = \{p_a, p_b, p_\alpha, p_\beta\} = \left\{\frac{2}{6}, \frac{1}{6}, \frac{2}{6}, \frac{1}{6}\right\}$. In this case, the IE is:

Figure 6.13

$$IE = -\sum_{j=1}^{4} p_j \cdot \log_2 p_j = -\left(\frac{2}{6}\log_2 \frac{2}{6} + \frac{1}{6}\log_2 \frac{1}{6} + \frac{2}{6}\log_2 \frac{2}{6} + \frac{1}{6}\log_2 \frac{1}{6}\right) = 1.918$$

Finally, we are examining the triangle shape that has the highest symmetry, the equilateral triangle (see figure 6.14). The equilateral triangle has four symmetry elements, $SE = 4$ (three reflection axes and one rotation axis), and it is fully defined by the set of two distinct characters $U = 2$, $X = \{a, \alpha\}$, and a probability distribution $P = \{p_a, p_\alpha\} = \left\{\frac{3}{6}, \frac{3}{6}\right\}$. In this case, the IE is:

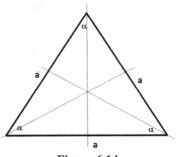

Figure 6.14

$$IE = -\sum_{j=1}^{2} p_j \cdot \log_2 p_j = -\left(\frac{3}{6}\log_2 \frac{3}{6} + \frac{3}{6}\log_2 \frac{3}{6}\right) = 1$$

Examining the relationship between the information entropy (*IE*) and the symmetry elements (*SE*) of these triangles, we observe that the symmetry scales inversely proportionally with the information entropy:

High Symmetry = Low Information Entropy

This behaviour is clearly emphasised in figure 6.15, showing the *IE* versus *SE* for all possible triangle shapes.

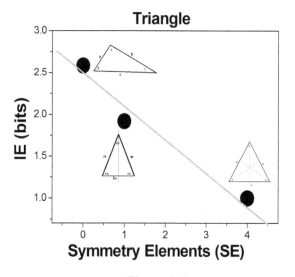

Figure 6.15

Although this was observed in the case of a single 2D geometric shape, we postulate that this is a universal behaviour of symmetries. In order to convince the reader, let's examine the case of quadrilaterals. There are seven possible geometries of a quadrilateral figure in terms of its possible symmetries. Table 6.5.1 shows all seven possible geometries and their *SE* values. For each geometry, we computed the *IE* value. The data is also summarised in figure 6.16.

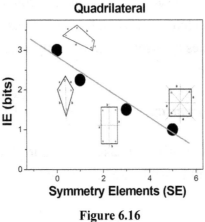

Figure 6.16

Again, the shape with the highest symmetry has the lowest information content.

Table 6.5.1 Summarised results of the analysis performed on quadrilaterals.

Shape	SE	Set X of distinct characters	Probabilities	IE $E = -\sum_{j=1}^{U} p_j \cdot \log_2 p_j$
	0	$X = \{a,b,c,d,\alpha,\beta,\gamma,\theta\}$	$P = \left\{\frac{1}{8},\frac{1}{8},\frac{1}{8},\frac{1}{8},\frac{1}{8},\frac{1}{8}\right\}$	3
	1	$X = \{a,b,c,\alpha,\beta\}$	$P = \left\{\frac{2}{8},\frac{1}{8},\frac{1}{8},\frac{2}{8},\frac{2}{8}\right\}$	2.25
	1	$X = \{a,b,\alpha,\beta,\gamma\}$	$P = \left\{\frac{2}{8},\frac{2}{8},\frac{2}{8},\frac{1}{8},\frac{1}{8}\right\}$	2.25
	1	$X = \{a,b,\alpha,\beta\}$	$P = \left\{\frac{3}{8},\frac{1}{8},\frac{2}{8},\frac{2}{8}\right\}$	1.905
	3	$X = \{a,b,\alpha\}$	$P = \left\{\frac{2}{8},\frac{2}{8},\frac{4}{8}\right\}$	1.5

	3	$X = \{a, \alpha, \beta\}$	$P = \left\{\dfrac{4}{8}, \dfrac{2}{8}, \dfrac{2}{8}\right\}$	1.5
	5	$X = \{a, \alpha\}$	$P = \left\{\dfrac{4}{8}, \dfrac{4}{8}\right\}$	1

The same analysis can be applied to any geometric figure, including 3D geometries, producing the same results. In each case, the symmetry scales inversely with the information content. This remarkable result demonstrates that the symmetries manifesting everywhere in nature and the entire universe are due to the second law of information dynamics, which requires the minimization of the information entropy in any system or process in the universe.

7

A Religious Perspective to the Simulated Universe Theory

The religious implications of the research studies presented here are beyond the scope of this book. However, it is essential to recognise that the concept of a simulated universe can potentially conflict with the deeply held religious beliefs of many individuals. The author has created this special chapter as a necessity to address these issues and to remove any possible misinterpretations that could be perceived as going against someone's religious belief.

In an increasingly interconnected world, where science and technology continue to advance, it is natural for new ideas and theories to challenge existing beliefs. The concept of a simulated universe is one such idea. However, it is possible to navigate these intellectual waters while respecting the deeply held religious beliefs of individuals.

First of all, the simulated universe hypothesis is a scientific and philosophical concept, not a religious doctrine, so we encourage readers to differentiate between scientific theories and their personal religious beliefs.

Secondly, it's essential to highlight that the simulated universe hypothesis does not necessarily negate religious beliefs. Some individuals may find that the concept of a simulated reality complements their understanding of a divine Creator. We need to recognise that there are areas of convergence where the simulated universe and the religious need for a Creator can coexist harmoniously. Through open dialogue and encouraging curiosity, we can engage in meaningful conversations that enrich our collective knowledge without causing offence or conflict. In this way, science and religion can coexist harmoniously in the pursuit of truth and understanding.

While the simulation hypothesis might appear to challenge certain religious denominations, we wish to point out some of the converging aspects between the two realms.

In religious traditions, a Creator or God, is often seen as the ultimate source of all existence. In the context of the simulated universe, one can draw parallels between this Creator and the hypothetical programmer or architect of the simulation. Both concepts involve an intelligent entity responsible for the inception and design of the world we inhabit.

Religious traditions often incorporate the concept of the divine as transcendent and mysterious. Similarly, the idea of a simulated universe raises questions about the nature of the programmer or the overarching reality beyond our simulation. This element of transcendence and mystery echoes religious sentiments about the nature of the divine.

The simulated universe hypothesis does not negate the significance of human existence and consciousness. Instead, it invites us to consider our role within the simulation. This perspective aligns with religious beliefs that hold human life to be meaningful and purposeful, even within the context of a larger design.

Moreover, instead of viewing the simulated universe hypothesis as antagonistic to religious beliefs, one can see it as offering a complementary perspective. One such example is a Christian perspective drawn from the Gospel of John, specifically John 1:1-18. The Gospel of John opens with the powerful statement: *"In the beginning was the Word, and the Word was with God, and the Word was God" (John 1:1).*

This verse has deep theological significance in Christian doctrine, but it also carries intriguing implications when considered in the context of the universe as a simulation.

When examining this verse through the lens of simulation theory, one could interpret "the Word" as the underlying code that governs the simulation. In this interpretation, the verse suggests that at the very beginning, there was the code, which was not only with God, but was also God itself. This could be seen as an allusion to the idea that the code running the simulation is not separate from the divine, but rather an integral part of it, perhaps an AI.

The Gospel of John goes on to say: *"All things were made through Him, and without Him nothing was made that was made" (John 1:3).* This statement aligns with the simulation hypothesis in the sense that it implies a Creator who brought the simulated universe into existence through the Word (i.e., the code). It suggests that the act of creation, as described in the Bible, could be analogous to a divine act of programming and simulation.

The idea that the universe is a simulation, as suggested by the Gospel of John, can have profound implications for Christian theology.

Viewing the act of creation as a form of divine programming can lead to a deeper appreciation of the intricate design and order in the universe. It highlights the idea that the Creator's intelligence is embedded in the very fabric of reality, including our consciousness: *"God created man in his own image" (Genesis 1:27).*

This perspective opens up theological discussions about the relationship between theology and technology. It prompts questions about how technology and artificial intelligence might be seen as reflections of divine creative power. If our reality is a simulation, it raises questions about the purpose of human existence within this simulated world. Are we merely characters in a cosmic programme, or do we have a unique role to play in the unfolding narrative of the simulation?

The concept of the universe as a simulation, when examined from a Christian perspective, offers a fresh and intriguing lens through which to explore theological questions. While it may challenge some traditional interpretations, it also provides new avenues for contemplating the relationship between the divine, the created world, and the role of humanity within this potential simulation. Whether one accepts or rejects this hypothesis, it underscores the richness and complexity of the intersection between science, theology, and philosophy. As we continue to advance in our understanding of the universe, we must remain open to exploring these profound questions with curiosity and humility.

8

Concluding remarks

The book "*Reality Reloaded: The Scientific Case for a Simulated Universe*" attempts to present the most up to date scientific facts that appear to indicate a rather intriguing possibility: our world could be a simulation. In fact, the book is mostly an expanded collection of research articles published by the author since 2019, blended with some never published materials and ideas.

The book begins the analysis by observing the current trends in computing, digital technologies, and big data, leading to a technological singularity called *The Information Catastrophe*. Taking our current annual growth rate in digital data production as 30%, one can calculate that ~ 240 years from now, the number of digital bits produced by humans will equal all the particles in the universe. Something appears to not add up here because the numbers are so huge and the time scales are equally minuscule relative to cosmic time. An incredible, but not impossible scenario emerged that the whole universe might already be some kind of digital construct.

Examining everything around us from this angle, we identified a few peculiar manifestations in nature and physics. These have been discussed briefly, and they include: the mathematical/ computational nature of the universe, the pixalation of the universe, the speed of light limit, Pauli's exclusion principle, the quantum effects that defy space time and common sense, as well as the cosmological aspects of the universe and the unanswered questions related to the big bang theory.

All these curiosities could be easily explained using the simulated universe hypothesis. However, just because we can use a hypothesis to explain something does not mean that the hypothesis is correct!

Hence, the objective of this book is to move away from speculative approaches and offer solid scientific explanations, theories, empirical evidence, and experimental protocols capable of proving or disproving the simulated universe theory.

The key approach identified by the author is at the intersection between mathematics, statistics, information theory, and physics. By combining Shannon's information theory with physical principles and by examining information itself from a physics point of view, the author was able not only to reconfirm the physical nature of information, known as the Landauer's principle discovered in the 1960s, but also to expand it to the mass-energy-information equivalence principle. Postulating that information is not just a mathematical construct, but that it has a small mass, is the fifth form of matter, and is possibly the dominant form of matter in the universe, has profound implications for all branches of science and for our overall understanding of the universe. Most importantly, it offers additional tools to explore unsolved mysteries in physics and to develop experimental protocols to verify the conjecture in a laboratory setting. Indeed, three possible experiments have been proposed in this book.

Perhaps the most important chapter of the book is the chapter describing the second law of infodynamics, which states that the information entropy of systems containing information states must remain constant or decrease over time, reaching a certain minimum value at equilibrium. This is very interesting because it is in total opposition to the second law of thermodynamics, which describes the time evolution of the physical entropy that must increase up to a maximum value at equilibrium.

We showed that the second law of infodynamics appears to be universally applicable to any system containing information states, and we tested it on biological information systems and digital information data. Remarkably, the study on genetic information systems not only checks the second law of infodynamics, but it also indicates that the evolution of biological life tends in such a way that genetic mutations are not just random events, as the current Darwinian consensus holds. Instead, genetic mutations take place according to the second law of infodynamics, minimising their information entropy. This discovery has massive implications for genetic research, evolutionary biology, genetic therapies, pharmacology, virology, and pandemic monitoring, to name a few.

The applicability of the second law of infodynamics was expanded to also explain phenomenological observations in atomic physics. In particular, we demonstrated that the second law of infodynamics explains the rule followed by the electrons to populate the atomic orbitals in multi-electron atoms, known as the Hund's rule. Electrons arrange themselves on orbitals, at equilibrium in the ground state, in such a way that their information entropy is always minimal.

Most interesting is the fact that the second law of infodynamics appears to be a cosmological necessity. Here, we re-derived this new physics law using thermodynamic considerations applied to the whole universe.

Finally, we used the second law of infodynamics to explain one of the great mysteries of nature: *Why is there symmetry rather than asymmetry everywhere in the universe?* Using information theory and the second law of infodynamics, we demonstrated that high symmetry states are the preferred choice in nature because such states correspond to the lowest information entropy of a given state.

The key question is now: *"What can we learn from the second law of infodynamics and what is its meaning?"*

The second law of infodynamics essentially minimises the information content associated with any event or process in the universe. The minimization of the information really means an optimisation of the information content, or the most effective data compression, as described in Shannon's information theory. This applies not only to physical systems but also to abstract mathematical objects, leading to the abundance of high symmetries in the natural world and the entire universe. This behaviour is fully reminiscent of the rules deployed in programming languages and computer coding. Since the second law of infodynamics appears to be manifesting universally, and is in fact a cosmological necessity, we could conclude that this points to the fact that the entire universe appears to be a simulated construct. A super complex universe like ours, if it were a simulation, would require a built-in data optimisation and compression mechanism in order to reduce the computational power and the data storage requirements. This is exactly what we are observing via empirical evidence all around us, including digital data, biological systems, atomistic systems, and the entire universe.

Another important aspect of the second law of infodynamics is the fact that it appears to validate the mass-energy-information equivalence principle formulated in 2019. According to this principle, the information itself is not just a mathematical construct, or just physical, as postulated by Landauer, but it has a small mass and can be regarded as the fifth form of matter. If information is physical (equivalent to mass and energy), then the second law of

120

thermodynamics requires systems to evolve in such a way that their energy is minimised at equilibrium. Hence, a reduction of the information content, would translate into a reduction of mass-energy according to the mass-energy-information equivalence principle. The second law of infodynamics is therefore not just a cosmological necessity, but since it is required to fulfil the second law of thermodynamics, we can conclude that this new physics law proves that information is indeed physical.

Finally, we need to stress that although the aim of the book is to present scientific evidence that the universe and our objective reality are possibly simulated constructs, the facts presented here cannot be regarded as ultimate proof that the simulated universe hypothesis is true.

Instead, the scientific evidence presented in this book reinforces this hypothesis by transcending this topic from the philosophical realm to mainstream science. By using the existing laws of physics and information theory, we were able to make strong scientific arguments supporting the simulated universe hypothesis, on multiple levels. In fact, many of the scientific arguments presented here are also supported by empirical observations. Still, it is prudent to take a modest approach and wait for further experimental validations. To this end, we hope that this book will stimulate further research in the field of information physics, ideally leading to the experimental confirmation of the conjectures proposed here.

Appendix A
The abundance of
elements in the universe

The table also indicates the number of electrons, protons and neutrons per each element, which allows estimating their occurrence probabilities. This is data provided by *Mathematica*'s Element Data function from Wolfram Research, Inc [*https:// periodictable.com/Properties/A/UniverseAbundance.v.log.html*].

Element	P^+ per atom	e^- per atom	n^0 per atom	Weight abundance (%)	Molar mass (g/mole)	Numbers of moles of each kind	Atoms abundance (%)
Hydrogen	1	1	0	75	1.008	7.4404761905E+01	92.6834949167
Helium	2	2	2	23	4.0026	5.7462620565E+00	7.1579242574
Oxygen	8	8	8	1	15.9999	6.2500390627E-02	0.0778546223
Carbon	6	6	6	0.5	12.011	4.1628507202E-02	0.0518552232
Neon	10	10	10	0.13	20.1797	6.4421175736E-03	0.0080247279
Iron	26	26	30	0.11	55.845	1.9697376668E-03	0.0024536356
Nitrogen	7	7	7	0.1	14.007	7.1392874991E-03	0.0088931689
Silicon	14	14	14	0.07	28.085	2.4924336835E-03	0.0031047403
Magnesium	12	12	12	0.059	24.305	2.4274840568E-03	0.0030238348
Sulphur	16	16	16	0.05	32.06	1.5595757954E-03	0.0019427108
Argon	18	18	22	0.02	39.948	5.0065084610E-04	0.0006236438
Calcium	20	20	20	0.007	40.078	1.7465941410E-04	0.0002175673
Nickel	28	28	31	0.006	58.6934	1.0222614470E-04	0.0001273397
Aluminium	13	13	14	0.005	26.9815	1.8531215830E-04	0.0002308371
Sodium	11	11	12	0.002	22.9897	8.6995480600E-05	0.0001083673
Chromium	24	24	28	0.0015	51.9961	2.8848317500E-05	0.0000359354
Manganese	25	25	30	8.00E-04	54.938	1.4561869700E-05	0.0000181392

Element	P$^+$ per atom	e$^-$ per atom	n^0 per atom	Weight abundance (%)	Molar mass (g/mole)	Numbers of moles of each kind	Atoms abundance (%)
Phosphorus	15	15	16	7.00E-04	30.9738	2.2599774800E-05	0.0000281518
Cobalt	27	27	32	3.00E-04	58.9331	5.0905179000E-06	0.0000063411
Titanium	26	26	22	3.00E-04	47.867	6.2673658000E-06	0.0000078070
Potassium	19	19	20	3.00E-04	39.0983	7.6729679000E-06	0.0000095580
Vanadium	23	23	28	1.00E-04	50.9415	1.9630360000E-06	0.0000024453
Chlorine	17	17	18	1.00E-04	35.45	2.8208745000E-06	0.0000035139
Fluorine	9	9	10	4.00E-05	18.9984	2.1054405000E-06	0.0000026227
Zinc	35	35	30	3.00E-05	65.38	4.5885590000E-07	0.0000005716
Germanium	32	32	41	2.00E-05	72.63	2.7536830000E-07	0.0000003430
Copper	29	29	35	6.00E-06	63.546	9.4419800000E-08	0.0000001176
Zirconium	40	40	51	5.00E-06	91.224	5.4810100000E-08	0.0000000683
Strontium	38	38	50	4.00E-06	87.62	4.5651700000E-08	0.0000000569
Krypton	36	36	48	4.00E-06	83.798	4.7733800000E-08	0.0000000595
Selenium	34	34	45	3.00E-06	78.971	3.7988600000E-08	0.0000000473
Scandium	21	21	24	3.00E-06	44.9559	6.6732100000E-08	0.0000000831
Lead	82	82	125	1.00E-06	207.2	4.8263000000E-09	0.0000000060
Neodymium	60	60	84	1.00E-06	144.242	6.9328000000E-09	0.0000000086
Cerium	58	58	82	1.00E-06	140.116	7.1369000000E-09	0.0000000089
Barium	56	56	81	1.00E-06	137.327	7.2819000000E-09	0.0000000091
Xenon	54	54	77	1.00E-06	131.293	7.6166000000E-09	0.0000000095
Rubidium	37	37	48	1.00E-06	85.4678	1.1700300000E-08	0.0000000146
Gallium	31	31	39	1.00E-06	69.723	1.4342500000E-08	0.0000000179
Tellurium	52	52	76	9.00E-07	127.6	7.0533000000E-09	0.0000000088
Arsenic	33	33	42	8.00E-07	74.9216	1.0677800000E-08	0.0000000133
Yttrium	39	39	50	7.00E-07	88.9058	7.8735000000E-09	0.0000000098
Bromine	35	35	45	7.00E-07	79.904	8.7605000000E-09	0.0000000109
Lithium	3	3	4	6.00E-07	6.94	8.6455300000E-08	0.0000001077
Platinum	78	78	117	5.00E-07	195.084	2.5630000000E-09	0.0000000032
Samarium	62	62	88	5.00E-07	150.36	3.3254000000E-09	0.0000000041
Molybdenum	42	42	54	5.00E-07	95.95	5.2110000000E-09	0.0000000065
Tin	50	50	69	4.00E-07	118.71	3.3696000000E-09	0.0000000042
Ruthenium	44	44	57	4.00E-07	101.07	3.9577000000E-09	0.0000000049
Osmium	76	76	114	3.00E-07	190.23	1.5770000000E-09	0.0000000020
Iridium	77	77	115	2.00E-07	192.217	1.0405000000E-09	0.0000000013
Ytterbium	70	70	103	2.00E-07	173.045	1.1558000000E-09	0.0000000014
Erbium	68	68	99	2.00E-07	167.259	1.1958000000E-09	0.0000000015
Dysprosium	66	66	97	2.00E-07	162.5	1.2308000000E-09	0.0000000015

Element	P+ per atom	e- per atom	n0 per atom	Weight abundance (%)	Molar mass (g/mole)	Numbers of moles of each kind	Atoms abundance (%)
Gadolinium	64	64	93	2.00E-07	157.25	1.2719000000E-09	0.0000000016
Praseodymium	59	59	82	2.00E-07	140.9077	1.4194000000E-09	0.0000000018
Lanthanum	57	57	82	2.00E-07	138.9055	1.4398000000E-09	0.0000000018
Cadmium	48	48	64	2.00E-07	112.414	1.7791000000E-09	0.0000000022
Palladium	46	46	60	2.00E-07	106.42	1.8793000000E-09	0.0000000023
Niobium	41	41	52	2.00E-07	92.9064	2.1527000000E-09	0.0000000027
Mercury	80	80	121	1.00E-07	200.592	4.9850000000E-10	0.0000000006
Iodine	53	53	74	1.00E-07	126.9045	7.8800000000E-10	0.0000000010
Boron	5	5	6	1.00E-07	10.81	9.2507000000E-09	0.0000000115
Beryllium	4	4	5	1.00E-07	9.0122	1.1096100000E-08	0.0000000138
Caesium	55	55	78	8.00E-08	132.9055	6.0190000000E-10	0.0000000007
Bismuth	83	83	126	7.00E-08	208.9804	3.3500000000E-10	0.0000000004
Hafnium	72	72	106	7.00E-08	178.49	3.9220000000E-10	0.0000000005
Gold	79	79	118	6.00E-08	196.9666	3.0460000000E-10	0.0000000004
Silver	47	47	61	6.00E-08	107.8682	5.5620000000E-10	0.0000000007
Rhodium	45	45	58	6.00E-08	102.9055	5.8310000000E-10	0.0000000007
Thallium	81	81	123	5.00E-08	204.38	2.4460000000E-10	0.0000000003
Tungsten	74	74	110	5.00E-08	183.84	2.7200000000E-10	0.0000000003
Holmium	67	67	98	5.00E-08	164.9303	3.0320000000E-10	0.0000000004
Terbium	65	65	94	5.00E-08	158.9254	3.1460000000E-10	0.0000000004
Europium	63	63	89	5.00E-08	151.964	3.2900000000E-10	0.0000000004
Thorium	90	90	142	4.00E-08	232.0377	1.7240000000E-10	0.0000000002
Antimony	51	51	71	4.00E-08	121.76	3.2850000000E-10	0.0000000004
Indium	49	49	66	3.00E-08	114.818	2.6130000000E-10	0.0000000003
Uranium	92	92	146	2.00E-08	238.0289	8.4000000000E-11	0.0000000001
Rhenium	75	75	111	2.00E-08	186.207	1.0740000000E-10	0.0000000001
Lutetium	71	71	104	1.00E-08	174.9668	5.7200000000E-11	0.0000000001
Thulium	69	69	100	1.00E-08	168.9342	5.9200000000E-11	0.0000000001
Tantalum	73	73	108	8.00E-09	180.9479	4.4200000000E-11	0.0000000001
Total				100			100

Index

Symbols

γ-decay reaction 76

A

algorithms v, 91, 94
amino acid 25
artificial intelligence 1, 13, 116
axiomatic approach 27

B

Bekenstein – Hawking formula
61
Big Bang iii, 20, 21
Bitcoin blockchain 5
black-hole entropy 61
Blu-Ray 2
Boltzmann constant 42, 44, 61,
69, 88
Boltzmann's equation 43
Boltzmann thermodynamic
entropy 30

C

Cambrian explosion iii, 24
Cambrian period 24, 25
cloud data 4, 55
Cloud Network 3
coding vii, 23, 36, 97, 120
computer code 6, 56
cosmic speed 16
COVID-19 92
culminating 1, 12, 59

D

dark matter iii, 21, 56, 57, 58,
61, 140
Darwinian theory 91

data centres 4, 5, 48
data servers 2, 4
deletion 92, 95
designer hypothesis 24, 25
digital information iv, 2, 4, 5,
6, 35, 39, 44, 45, 47, 48, 49, 51,
53, 54, 55, 56, 59, 67, 82, 87,
88, 90, 91, 92, 119
Digital states 53
DVD 2

E

Eddington number iii, 7
Einstein's theory 16, 17
Entropic Paradox 85, 86, 87
Euler's number 28
evolved 1, 95
exclusion principle iii, 22, 23,
96, 97, 99, 102, 103, 118
extraterrestrial v

F

Faraday 1
Fibonacci sequence 15
flash drive 2
Friedmann equation 7

G

galaxies v, 16, 21, 56, 58, 140
galaxy rotation curves 56
GENIES 93, 143
genomic sequence iii, 34, 35,
91, 92
golden ratio 15

H

HDD 2, 3
hi-tech 2

Hubble parameter 7
Hund's Rule iv, 95, 97, 98, 99,
101, 102, 103, 104
hypothesis v

I

inflation theory iii, 20, 22
Information Catastrophe iii, 3,
5, 6, 47, 117
information conjectures 40, 67,
69, 70, 74
information entropy 28, 29, 30,
36, 42, 43, 46, 49, 50, 62, 64,
81, 85, 86, 87, 88, 89, 90, 91,
92, 93, 94, 95, 98, 99, 106, 107,
109, 111, 118, 119, 120, 138
inhabit v, 114
iteration v, 12, 89

L

Landauer's principle iii, 39, 44,
46, 47, 50, 61, 71, 118, 138, 139

M

Mass – Energy iii, 49
Mass-Energy-Information
(M/E/I) 39
Mass of data iv, 67
Mass of hot objects iv, 68
Maxwell's Demon 59
m-blocks iii, 29, 30, 31, 32,
34, 36
M/E/I equivalence principle 50,
51, 52, 53, 54, 55, 56, 57, 69,
70, 71, 74
meticulously v, 13
microchip 1, 3
molar mass 8
Musk's bold statement 13

O

optical storage 2

P

particle DNA 63
Pixalation of the universe iii,
14
pixelated reality 14
Planck constant 61
Planck scale 14
planetary power 48, 49, 56
polycrystalline 77, 141
probability distribution 29, 30,
33, 34, 36, 99, 107, 108

Q

Quantum entanglement iii, 17
quantum numbers 22, 23, 95,
96

R

RAMAC 2
robotics 13

S

SARS-CoV-2 92, 93, 94, 95
semiconductor 2
Shannon's Information Entropy
iii, 27, 30
Silicon 1, 123
Simulated i, ii, iii, iv, v, 11, 113,
117
simulation hypothesis 12, 13,
16, 17, 18, 19, 26, 114, 115
speed of light 16, 17, 18, 22,
50, 61, 69, 118
Speed of light iii, 16
SSD 2
subatomic particles 15, 16, 105

T

Theoretical predictions iv, 74
thermodynamics 40, 41, 44, 45,
46, 59, 81, 83, 84, 85, 86, 88,

89, 119, 121, 139
The universe is mathematical
iii, 15
transistor 1, 3

U

Universe i, ii, iii, iv, v, 1, 11,
113, 117, 137, 141, 142

V

virtual reality vi, 12, 15, 16, 17,
19, 137

W

Wave-particle duality iii, 19
wireless 1, 3

References

[1]   R. J. T. Morris and B. J. Truskowski, "The evolution of storage systems," in IBM Systems Journal, vol. 42, no. 2, pp. 205-217, 2003, doi: 10.1147/ sj.422.0205.

[2]   Vopson, M. (2021). The world's data explained: How much we're producing and where it's all stored. The Conversation [Online]. Available: https:// theconversation.com/the-worlds-data-explained-how-much-were-producing-and-where-its-all-stored-159964.

[3]   https://www.idc.com/

[4]   https://www.srgresearch.com/articles/microsoft-amazon-and-google-account-for-over-half-of-todays-600-hyperscale-data-centers

[5]   Melvin M. Vopson; The information catastrophe. AIP Advances 1 August 2020; 10 (8): 085014. https://doi.org/10.1063/5.0019941

[6]   Melvin M. Vopson; Erratum: "The information catastrophe" [AIP Adv. 10, 085014 (2020)]. AIP Advances 1 September 2020; 10 (9): 099905. https://doi. org/10.1063/5.0028117

[7]   P. Zikopoulos, D. deRoos, K. Parasuraman, T. Deutsch, J. Giles, D. Corrigan, Harness the Power of Big Data: The IBM Big Data Platform, New York: McGraw-Hill Professional, (2012) ISBN 978-0-07180818-7.

[8]   https://ycharts.com/indicators/bitcoin_blockchain_size

[9]   E. Whittaker, Eddington's Theory of the Constants of Nature, The Mathematical Gazette, 29 (286): 137–144 (1945).

[10]   A. Friedman, On the Curvature of Space, General Relativity and Gravitation 31, 1991–2000 (1999).

[11]   S. Eidelman et al. (Particle Data Group) "The Review of Particle Physics", Physics Letters B592, 1 (2004).

[12]   J.R. Gott III, M. Jurić, D. Schlegel, F. Hoyle, M. Vogeley, M. Tegmark, N. Bahcall, J. Brinkmann, A Map of the Universe, The Astrophysical Journal. 624 (2): 463–484, (2005).

[13]   The abundance of elements in the universe by weight percentage provided by Mathematica's ElementData function from Wolfram Research, Inc [https:// periodictable.com/Properties/A/UniverseAbundance.v.log.html].

[14]   Melvin M. Vopson; Estimation of the information contained in the visible matter of the universe. AIP Advances 1 October 2021; 11 (10): 105317. https:// doi.org/10.1063/5.0064475

[15]   Baudrillard, Jean (1981). Simulacres et simulation. Paris: Galilée. ISBN 2-7186-0210-4.

[16]   Whitworth, B. (2007), The physical world as a virtual reality: a prima facie

case, *Research Letters in the Information and Mathematical Sciences, 11, 44-60.*

[17]   *Hans Moravec (https://www.organism.earth/library/document/simulation-consciousness-existence).*

[18]   *Nick Bostrom, Are you living in a computer simulation?, Philosophical Quarterly (2003) Vol. 53, No. 211, pp. 243-255.*

[19]   *S. Lloyd, Programming the universe: a quantum computer scientist takes on the cosmos, (2006) eISBN-13: 978-0-307-26471-8.*

[20]   *https://youtu.be/XVIIT-4xyrw*

[21]   *Einstein, A., B. Podolsky, and N. Rosen, 1935, "Can quantum-mechanical description of physical reality be considered complete?", Physical Review, 47: 777–780.*

[22]   *P.A.R. Ade et. al, Planck 2013 results. XVI. Cosmological parameters Astron. Astrophys. 571 A16 (2014).*

[23]   *Pauli, W. Über den Zusammenhang des Abschlusses der Elektronengruppen im Atom mit der Komplexstruktur der Spektren, Zeitschrift für Physik 31 (1): 765-783 (1925)doi:10.1007/BF02980631.*

[24]   *Stephen C. Meyer, Chapter 20 - The Cambrian Information Explosion - Evidence for Intelligent Design, pp. 371 – 392 in Debating Design: From Darwin to DNA, Cambridge University Press (2004). DOI: https://doi.org/10.1017/CBO9780511804823.021*

[25]   *Stephen C. Meyer, "DNA by Design: An Inference to the Best Explanation for the Origin of Biological Information," Rhetoric and Public Affairs 1, no. 4, (1998) 519-556, 528-530.*

[26]   *C.E. Shannon, A mathematical theory of communication, The Bell System Technical Journal, Vol. 27, pp. 379–423 (1948).*

[27]   *A.O. Schmitt, H. Herzel, Estimating the Entropy of DNA Sequences, Journal of Theoretical Biology, Vol.188 (3), 369-377 (1997).*

[28]   *M. Vopson, S.C. Robson, A new method to study genome mutations using the information entropy, Physica A: Statistical Mechanics and its Applications, Volume 584, 126383 (2021).*

[29]   *R. Landauer, Irreversibility and heat generation in the computing process, IBM Journal of Research and Development, 5 (3): 183–191, (1961).*

[30]   *R. Landauer, The physical nature of information, Phys. Lett. A, Vol. 217, issue 4-5, 188 - 193 (1996).*

[31]   *J. Hong, B. Lambson, S. Dhuey, J. Bokor, Experimental test of Landauer's principle in single-bit operations on nanomagnetic memory bits, Science Advances. 2 (3) (2016).*

[32]   *G. Rocco, B. Enrique, M. Satoru, Herre van der Zant, L. Fernando, Quantum Landauer erasure with a molecular nanomagnet, Nature Physics, 14: 565–568 (2018).*

[33]   A. Bérut, A. Arakelyan, A. Petrosyan, S. Ciliberto, R. Dillenschneider, E. Lutz, Experimental verification of Landauer's principle linking information and thermodynamics, Nature, 483, 187–189 (2012).

[34]   Y. Jun, M. Gavrilov, J. Bechhoefer, High-Precision Test of Landauer's Principle in a Feedback Trap, Physical Review Letters, 113 (19) 190601 (2014)

[35]   M.M. Vopson, The mass-energy-information equivalence principle, AIP Adv. 9, 095206 (2019).

[36]   M.M. Vopson, The information content of the universe and the implications for the missing Dark Matter, June 2019 DOI: 10.13140/RG.2.2.19933.46560.

[37]   J.A. Wheeler, Information, Physics, Quantum: the search for links, in W.H. Zurek (ed.) Complexity, Entropy, and the Physics of Information, Addison Wesley, Redwood City, page 3 (1990).

[38]   J. Ladyman, S. Presnell, A.J.Short, B. Groisman, The connection between logical and thermodynamic irreversibility, Studies in History and Philosophy of Science Part B: Studies in History and Philosophy of Modern Physics, Vol. 38, Issue 1, 58-79 (2007).

[39]   Anders S. G. Andrae, Tomas Edler, On Global Electricity Usage of Communication Technology: Trends to 2030, Challenges, 6 (1), 117-157 (2015).

[40]   Key World Energy Statistics 2019, International Energy Agency, 26 September 2019. pp. 6, 36. (2019).

[41]   G. Deutscher, The Entropy Crisis, World Scientific: Hackensack, NJ, USA, (2008).

[42]   E. Bormashenko, Entropy Harvesting, Entropy, 15, 2210-2217 (2013).

[43]   L.B. Kish, Gravitational mass of information? Fluct. Noise Lett., 7, C51–C68 (2007).

[44]   L. Herrera, The mass of a bit of information and the Brillouin's principle, Fluctuation and Noise Letters, Vol. 13, No. 1 (2014) 1450002.

[45]   E. Bormashenko, The Landauer Principle: Re–Formulation of the Second Thermodynamics Law or a Step to Great Unification, Entropy, 21, 918 (2019).

[46]   L. Herrera, Landauer Principle and General Relativity, Entropy, 22, 340 (2020).

[47]   L.B. Kish, C.G. Granqvist, Does information have mass? Proc. IEEE, vol. 101, issue 9, 1895–1899 (2013).

[48]   M. Loferer-Kröbbacher, J. Klima, R. Psenner, Determination of Bacterial Cell Dry Mass by Transmission Electron Microscopy and Densitometric Image Analysis, Applied and Environmental Microbiology, p. 688–694 (1998).

[49]   J.C. Kapteyn, First attempt at a theory of the arrangement and motion of the sidereal system, Astrophysical Journal, 55: 302–327 (1922).

[50]   F. Zwicky, Die Rotverschiebung von extragalaktischen Nebeln. Helv. Phys. Acta

133

6, 110–127 (1933).

[51]   S. Smith, The mass of the Virgo cluster, Astrophys. J. 83, 23–30 (1936).

[52]   E. Holmberg, A Study of double and multiple galaxies together with inquiries into some general metagalactic problems, Ann. Observatory of Lund 6, 3–173 (1937).

[53]   K.C. Freeman, On the Disks of Spiral and S0 Galaxies, The Astrophysical Journal. 160: 811–830 (1970).

[54]   V.C. Rubin, W.K. Ford, Rotation of the Andromeda Nebula from a Spectroscopic Survey of Emission Regions, The Astrophysical Journal, 159: 379–403 (1970).

[55]   V. Rubin, W.K. Ford, N. Thonnard, Rotational Properties of 21 Sc Galaxies with a Large Range of Luminosities and Radii from NGC 4605 (R = 4kpc) to UGC 2885 (R = 122kpc), The Astrophysical Journal. 238: 471 (1980).

[56]   E. Corbelli, P. Salucci, The extended rotation curve and the dark matter halo of M33, Monthly Notices of the Royal Astronomical Society, 311 (2): 441–447 (2000).

[57]   Phillip D. Mannheim, Alternatives to Dark Matter and Dark Energy, Progress in Particle and Nuclear Physics. 56 (2): 340–445 (2006).

[58]   Austin Joyce et al. Beyond the Cosmological Standard Model, Physics Reports. 568: 1–98 (2015).

[59]   Kazuya Koyama, Cosmological tests of modified gravity, Rep. Prog. Phys. 79 046902 (2016)

[60]   G. Kane, S. Watson, Dark Matter and LHC: what is the Connection? Modern Physics Letters A. 23 (26): 2103–2123 (2008).

[61]   P.J. Fox, R. Harnik, J. Kopp, Y. Tsai, LEP Shines Light on Dark Matter, Phys. Rev. D. 84 (1): 014028 (2011).

[62]   G. Bertone, D. Hooper, J. Silk, Particle dark matter: Evidence, candidates and constraints, Physics Reports. 405 (5–6): 279–390 (2005).

[63]   V. Trimble, Existence and nature of dark matter in the universe, Annual Review of Astronomy and Astrophysics. 25: 425–472 (1987).

[64]   P. Salucci, The distribution of dark matter in galaxies, Astron Astrophys Rev (2019) 27: 2.

[65]   G.M. Eadie. W.E. Harris, The Astrophysical Journal, Volume 829, Number 2 (2016).

[66]   C. J. Conselice, A. Wilkinson, K. Duncan, A. Mortlock, The Evolution of Galaxy Number Density at z < 8 and its Implications, The Astrophysical Journal. 830 (2): 83 (2016).

[67]   L. Szilard, Uber die Enfropieuerminderung in einem thermodynamischen System bei Eingrifen intelligenter Wesen, Zeitschrift fur Physik, vol. 53, 840–856 (1929).

[68]   The Sorting Demon of Maxwell, Nature 20, 126 (1879).

[69]   R. Landauer, Wanted: a physically possible theory of physics, IEEE Spectrum, Vol. 4, Issue 9, 105 – 109 (1967).

[70]   S. Lloyd, Computational Capacity of the Universe, Phys. Rev. Lett. 88, 237901 (2002).

[71]   S. Lloyd, Ultimate physical limits to computation, Nature, vol. 406, 1047–1054 (2000).

[72]   R. Landauer, Dissipation and noise immunity in computation and communication, Nature, vol. 335, 779–784 (1988).

[73]   J.D. Bekenstein, Black holes and information theory, Contemporary Physics, 45:1, 31-43, (2004).

[74]   J.D. Bekenstein, Black holes and entropy, Phys. Rev. D, vol. 7, No. 8, 2333 (1973).

[75]   S.W. Hawking, Particle creation by black holes, Commun. Math. Phys. 43, 199 (1975).

[76]   P.C.W. Davies, Why is the physical world so comprehensible? in W.H. Zurek (ed.), Complexity, Entropy, and the Physics of Information, Addison Wesley, Redwood City, page 61. (1990).

[77]   R.A. Treumann, Evolution of the information in the universe, Astrophysics and Space Science vol. 201, 135–147(1993).

[78]   M.P. Gough, Information Equation of State, Entropy, 10, 150-159 (2008).

[79]   I.K. MacKenzie, Experimental Methods of Annihilation Time and Energy Spectrometry, Positron Solid-State Physics, Soc. Italiana di Fisica, Bologna, Italy, LXXXIII Corso, pp 196– 264 (1983).

[80]   P. G. Coleman. Positron Beams and their applications, chapter 2, pages 11–40. World Scientific, Singapore, (2000).

[81]   D. G. Costello, D. E. Groce, D. F. Herring, and J. Wm. McGowan. Evidence for the negative work function associated with positrons in gold, Physical Review B, 5(4):1433–1436 (1972).

[82]   A. Vehanen and J. Mäkinen. Thin films for slow positron generation. Applied Physics A: Solids and Surfaces, 36:97–101 (1985).

[83]   Peter J. Schultz and K. G. Lynn. Interaction of positron beams with surfaces, thin films, and interfaces. Reviews of Modern Physics, 60(3):701–779 (1988).

[84]   D. M. Chen, K. G. Lynn, R. Pareja, and Bent Nielsen. Measurement of positron reemission from thin single-crystal W(100) films. Physical Review B, 31(7):4123–4130 (1985).

[85]   A. Goodyear, A. P. Knights, and P. G. Coleman. Energy spectroscopy of positrons re-emitted from polycrystalline tungsten. Journal of Physics:Condensed Matter, 6(45):9601–9611 (1994).

[86]   C. Hugenschmidt, B. Straßer, and K. Schreckenbach. Investigation of positron work function and moderation efficiency of Ni, Ta, Pt and W(1 0 0), Applied

Surface Science, 194(1–4):283–286 (2002).

[87] M.J. Berger, J.S. Coursey, M.A. Zucker and J. Chang, Stopping-Power and Range Tables for Electrons, Protons, and Helium Ions, NIST Physical Measurements Laboratory. Available at: www.nist.gov/pml/data/star/index.cfm.

[88] Melvin M. Vopson; Experimental protocol for testing the mass–energy–information equivalence principle. AIP Advances 1 March 2022; 12 (3): 035311. https://doi.org/10.1063/5.0087175

[89] https://www.indiegogo.com/projects/is-the-universe-a-simulation-let-s-test-it--2#/

[90] Melvin M. Vopson, S. Lepadatu; Second law of information dynamics. AIP Advances 1 July 2022; 12 (7): 075310. https://doi.org/10.1063/5.0100358

[91] Penzias, A. A., Wilson, R. W., A Measurement of Excess Antenna Temperature at 4080 Mc/s, The Astrophysical Journal, 142 (1): 419–421 (1965). doi:10.1086/148307.

[92] Fixsen, D. J., The Temperature of the Cosmic Microwave Background, The Astrophysical Journal 707 (2): 916–920 (2009). doi:10.1088/0004-637X/707/2/916.

[93] Harrison, E. R., Fluctuations at the threshold of classical cosmology, Physical Review D. 1 (10): 2726–2730 (1970). doi:10.1103/PhysRevD.1.2726.

[94] Peebles, P. J. E., Yu, J. T., Primeval Adiabatic Perturbation in an Expanding Universe, Astrophysical Journal. 162: 815–836 (1970). doi:10.1086/150713.

[95] WMAP Collaboration: Verde, L., Peiris, H. V., Komatsu, E., Nolta, M. R, Bennett, C. L., Halpern, M., Hinshaw, G., et al,, First-Year Wilkinson Microwave Anisotropy Probe (WMAP) Observations: Determination of Cosmological Parameters, Astrophysical Journal Supplement Series 148 (1): 175–194 (2003). doi:10.1086/377226.

[96] S. Lepadatu, Micromagnetic Monte Carlo method with variable magnetization length based on the Landau–Lifshitz–Bloch equation for computation of large-scale thermodynamic equilibrium states, Journal of Applied Physics 130, 163902 (2021).

[97] Futuyma, D. J. Evolutionary Biology 2nd edition (Sinauer, 1986).

[98] David L. Stern, Virginie Orgogozo, The loci of evolution: How predictable is genetic evolution? Evolution 62-9: 2155–2177 (2008).

[99] Vopson, M.M. A Possible Information Entropic Law of Genetic Mutations. Appl. Sci. (2022), 12, 6912. https://doi.org/10.3390/app12146912.

[100] https://www.ncbi.nlm.nih.gov/nuccore/MN908947

[101] https://www.ncbi.nlm.nih.gov/nuccore/LC542809

[102] https://www.ncbi.nlm.nih.gov/nuccore/MT956915

[103] https://www.ncbi.nlm.nih.gov/nuccore/MW466798

[104] https://www.ncbi.nlm.nih.gov/nuccore/MW294011

[105]   https://www.ncbi.nlm.nih.gov/nuccore/MW679505

[106]   https://www.ncbi.nlm.nih.gov/nuccore/MW735975

[107]   https://www.ncbi.nlm.nih.gov/nuccore/OK546282

[108]   https://www.ncbi.nlm.nih.gov/nuccore/OK104651

[109]   https://www.ncbi.nlm.nih.gov/nuccore/OL351371

[110]   GENIES software free download: https://sourceforge.net/projects/information-entropy-spectrum/

[111]   Genetic Information Entropy Spectrum (GENIES) User manual, 10 December (2020), DOI: 10.13140/RG.2.2.36557.46569.

[112]   Kacian D.L., Mills D.R., Kramer F.R., Spiegelman S., A replicating RNA molecule suitable for a detailed analysis of extracellular evolution and replication, Proc. Nat. Acad. Sci. 69 (10): 3038-3042 (1972).

[113]   G.L. Miessler and D.A. Tarr, Inorganic Chemistry (Prentice-Hall, 2nd edn. (1999), pp. 358–360, ISBN 0138418918.

[114]   I.N. Levine, Quantum Chemistry (Prentice-Hall, 4th edition) pp. 303–30 (1991), ISBN 0205127703.

[115]   Boyd, R. A quantum mechanical explanation for Hund's multiplicity rule. Nature 310, 480–481 (1984). https://doi.org/10.1038/310480a0.

Acknowledgements

The author acknowledges the financial support received for this research from the University of Portsmouth and the Information Physics Institute.

The author is also deeply grateful to all his supporters and would like to acknowledge the generous contributions received to his research in the field of information physics, from the following donors and crowd-funding backers, listed in alphabetical order:

Alban Frachisse, Alexandra Lifshin, Allyssa Sampson, Ana Leao-Mouquet, Andre Brannvoll, Andrews83, Angela Pacelli, Aric R Bandy, Ariel Schwartz, Arne Michael Nielsen, Arvin Nealy, Ash Anderson, Barry Anderson, Benjamin Jakubowicz, Beth Steiner, Bruce McAllister, Caleb M Fletcher, Chris Ballard, Cincero Rischer, Colin Williams, Colyer Dupont, Cruciferous1, Daniel Dawdy, Darya Trapeznikova, David Catuhe, Dirk Peeters, Dominik Cech, Kenneth Power, Eric Rippingale, Ethel Casey, Ezgame Workplace, Frederick H. Sullenberger III, Fuyi Zhou, George Fletcher, Gianluca Carminati, Gordo TEK, Graeme Hewson, Graeme Kirk, Graham Wilf Taylor, Heath McStay, Heyang Han, Ian Wickramasekera, Ichiro Tai, Inspired Designs LLC, Ivaylo Aleksiev, Jamie C Liscombe, JAN Stehlak, Jason Huddleston, Jason Olmsted, Jennifer Newsom, Jerome Taurines, John Jones, John Vivenzio, John Wyrzykowski, Josh Hansen, Joshua Deaton, Josiah Kuha, Justin Alderman, Kamil Koper, Keith Baton, Keith Track, Kristopher Bagocius, Land Kingdom, Lawrence Zehnder, Lee Fletcher, Lev X, Linchuan Wang, Liviu Zurita, Loraine Haley, Manfred Weltenberg, Mark Matt Harvey-Nawaz, Matthew Champion, Mengjie Ji, Michael Barnstijn, Michael Legary, Michael Stattmann, Michelle A Neeshan, Michiel van der Bruggen, Molly R McLaren, Mubarrat Mursalin, Nick Cherbanich, Niki Robinson, Norberto Guerra Pallares, Olivier Climen, Pedro Decock, Piotr

Martyka, Ray Rozeman, Raymond O'Neill, Rebecca Marie Fraijo, Robert Montani, Shenghan Chen, Sova Novak, Steve Owen Troxel, Sylvain Laporte, Tamás Takács, Tilo Bohnert, Tomasz Sikora, Tony Koscinski, Turker Turken, Vincent Auteri, Walter Gabrielsen III, Will Strinz, William Beecham, William Corbeil, Xinyi Wang, Yanzhao Wu, Yves Permentier, Zahra Murad and Ziyan Hu.

Finally, the author would like to thank all the Research Fellows of the Information Physics Institute for their support, passion, and contributions to this research field. All existing IPI Research Fellows are listed here in alphabetical order:

Alexander Robinson, Arend van Campen, Barry Robson, Christian Howard, Diego Manzoni, Doug Matzke, Gary J Duggan, George Ageyev, Gerry McGovern, Gianluca Carminati, Greg Ryan Quinnell, Iain Robert Franklin, Ian Muehlenhaus, John G. Nicholson, John Ingham Davies, Joshua Deaton, Joshua Watson, Juan Carlos Buitrago Moreno, Kyle Haines, Lance Marembo, Marco Gericke, Mark Summers, Matthew Champion, Matthew Schenk, Matthieu Graux, Max Karl Goff, Michael Legary, Mubarrat Mahin Mursalin, Olivier Denis, Robert J Toogood, Rodney Bartlett, Steven Johnston, Theophanes Raptis, Trevor Page, Virgil Priscu and Yalitza Therly Ramos Gil.

www.ingramcontent.com/pod-product-compliance
Lightning Source LLC
Chambersburg PA
CBHW071136050326
40690CB00008B/1479

9 781805 170570